ISBN 13: 978-3-662-16090-9 e-ISBN 13: 978-3-7985-1793-6

DOI: 10.1007/ 978-3-7985-1793-6

CONTENTS · INHALT

1. *Fitch, R. M.* and *Lih-bin Shih* (Connecticut/USA), Emulsion Polymerization: Kinetics of radial capture by the particles (With 8 figures and 5 tables) 1

2. *Forslind, E.* (Stockholm/Sweden), Liquid Water at Interfaces (With 2 figures and 2 tables) 12

3. *Friberg, S.* and *I. Lapczynska* (Stockholm/Sweden), Microemulsions and solubilization by nonionic surfactants (With 6 figures) 16

4. *Stenius, P.* and *L. A. Filén* (Åbo/Finland), The influence of solubilized N-Decanol on the stability of sodium N-Pentanoate (sodium valerate) micelles (With 6 figures and 3 tables) 21

5. *Svens, B.* and *M. Turpeinen* (Åbo/Finland), Investigation of isotropic micellar solutions in the system sodium octaanoate – N-Alcohol – water by small-angle X-Ray scattering With 5 figures and 1 table) 30

6. *Lincoln, Britta* (Stockholm/Sweden), The Solubilization of Water in Polyoxyethylene Alkyl Ethers and the Influence on the Bonds between the Ethers and Fatty Acids (With 9 figures) 37

PROGRESS IN COLLOID AND POLYMER SCIENCE

Supplements to "Colloid and Polymer Science" · Continuation of „Kolloid-Beihefte"

Vol. 56 1975

1

Department of Chemistry and Institute of Materials Science, University of Connecticut Storrs, Conn. 06268 (U.S.A.)

Emulsion polymerization: Kinetics of radical capture by the particles

R. M. Fitch and Lih-bin Shih

With 8 figures and 5 tables

(Received January 17, 1974)

Introduction

The rate at which free radicals enter particles of the disperse phase from the continuous phase in emulsion polymerization depends upon the particle concentration and size and the rate of radical generation. The kinetics of radical capture determine the rate of particle formation, relative growth rates of different sized particles in certain cases and the chemistry of the polymer/water interface. *Smith* and *Ewart* in their classic work (1, 2) assumed, on the basis of a suggestion by *P. Debye*, that the process of radical capture by particles was diffusion-controlled to calculate the particle number concentration and one part of their "Case 1" polymerization kinetics. Under diffusion control, the radical absorption rate is proportional to the radius of the particles. There is good evidence to support this model: *Ugelstad* and coworkers applied it to calculate relative particle growth rates (3) and overall kinetics of vinyl chloride emulsion polymerization (4), and *La Mer* and *Reiss* found that the growth of aerosols and hydrosols could be ascribed to diffusion-controlled absorption of small molecules by particles (5, 6). On the other hand *Gardon* proposed that the capture of radicals by monomer-swollen soap micelles should be described by a collision theory which gives a dependence of the capture rate on the square of the particle radius (7). *Fitch* and *Tsai* applied this theory to calculate particle nucleation kinetics (8). *Alexander* and *Napper* in an excellent review have recently stated that whether radical entry is diffusion- or collision-controlled is still an open, important question which needs to be resolved (9).

In this paper we investigate some of the assumptions of the original *Gardon* collision theory, and propose refinements to obviate them. Additionally we describe experiments using emulsion polymerization of methyl methacrylate to test which theory is the more applicable.

Collision theory

The rate at which radicals enter a given particle from the continuous medium, or the "capture rate" for a single particle, R_{C_1}, in terms of the number of radicals per second has been given as (7, 8):

$$R_{C_1} = \pi R_i r_p^2 L . \qquad [1]$$

This assumes that: (a) the polymer particle does not move, i.e. that its diffusion coefficient, D_p, need not be considered, (b) the size of the radicals is negligible compared to that of the particles, (c) the near presence of one particle does not interfere with the capture of radicals by others, i.e. there is no "crowding effect", (d) the system contains a uniform concentration of radicals throughout, and (e) there is no interaction between oligomers and particles. Each of these assumptions is discussed and the amended expression for the capture rate follows.

The probability and rate of capture may be visualized by reference to fig. 1. The polymer particle is represented by the solid circle of radius r_p, whose center is at B.

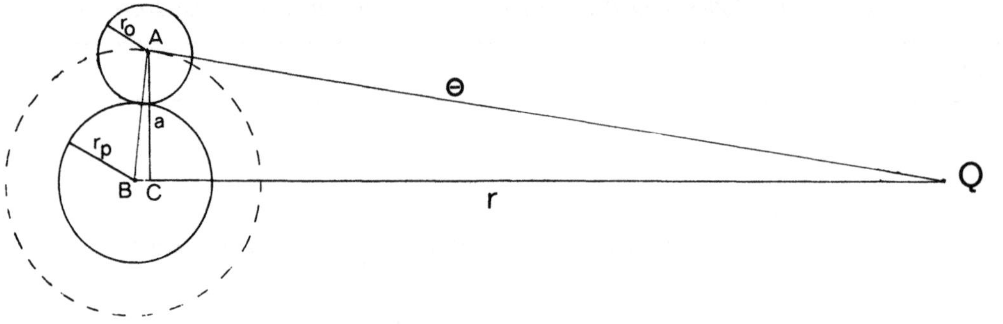

Fig. 1. Collision Probability Geometry

A free radical, which was initiated at point Q, in the solution will have a radius of gyration r_0 and its center at A by the time it has travelled the distance Θ, at which it comes into contact with the particle[1]).

The collision cross-section, A, offered to the radical when it is far away (r is large) is approximately defined by the sum of the radii r_p and r_0:

$$A = \pi(r_p + r_0)^2 . \quad [2]$$

At small r, the cross-section is better given by

$$A = \pi a^2 , \quad [3]$$

where a is the distance $\overline{AC}$ perpendicular to $\overline{BQ}$.

Because we shall be integrating over a relatively great distance along r, equation [2] may be employed with sufficient precision for our purposes. For similar reasons Θ may be taken as approximately equal to r, so that the collision probability, $P_{(r)}$, is

$$P_{(r)} = \frac{A}{4\pi\Theta^2} \approx \frac{\pi (r_p + r_0)^2}{4\pi r^2} . \quad [4]$$

The size of the oligomeric radical when it collides with the particle will depend upon the distance r. The dependence of r_0 on r is found as follows: The radius of gyration of a polymer molecule depends on the square root of the molecular weight (10):

$$r_0 = k_1 M_0^{0,5} . \quad [5]$$

The proportionality constant, k_1, is slightly dependent on molecular weight and the square root dependence holds strictly only for statistically large molecules, but these deviations are small and we shall neglect them. The molecular weight depends on the degree of polymerization, X:

$$M_0 = M_m X , \quad [6]$$

where M_m is the molecular weight of a monomer unit. The value of X increases with time, t, and the rate of chain growth:

$$X = k_p [M]_{\text{cont}} \cdot t , \quad [7]$$

where k_p is the propagation rate constant and $[M]_{\text{cont}}$ is the concentration of monomer in the aqueous phase. Since during a short period of interest $[M]_{\text{cont}}$ will remain sensibly constant, we see by combining equations [5], [6] and [7], that

$$r_0 = k_2 t^{\frac{1}{2}} ,$$

$$\text{where } k_2 = k_1 (M_m k_p [M]_{\text{cont}})^{\frac{1}{2}} . \quad [8]$$

The distance, r, the oligomer diffuses during the time t may be obtained from *Einstein's Brownian* motion equation,

$$t = \frac{r^2}{2 (D_p + \bar{D}_0)} , \quad [9]$$

in which $\bar{D}_0$ is the time-average diffusion coefficient for the growing radical. Combination of equations [9] and [8] and taking all the constant factors together gives the very simple dependence of the radius of gyration on the distance travelled:

$$r_0 = \bar{K} r,$$

$$\text{where } \bar{K} = k_1 \left(\frac{M_m k_p [M]_{\text{cont}}}{2 (D_p + \bar{D}_0)} \right)^{\frac{1}{2}} . \quad [10]$$

---

1) Θ is a tangent from Q and is the maximum distance. The minimum distance is $r - r_p - r_0$. The difference is considered negligible *(vide infra)*.

This, when substituted into equation [4], gives to a good approximation the capture probability, $P_{(r)}$, in which assumptions (a) and (b) have been removed:

$$P_{(r)} = \frac{(r_p + \bar{K}r)^2}{4r^2}. \qquad [11]$$

The rate of capture is found from the product of the probability, $P_{(r)}$, and the rate of generation of radicals, $\varrho_{(r)}$, at all values of r out to the limit $r = L + r_p + r_0$. Radicals formed at greater distances will self-nucleate before reaching the particle. At any distance r, the value of $\varrho_{(r)}$ will be the product of the overall rate of radical formation and the volume of the spherical shell surrounding the particle:

$$\varrho_{(r)} = R_i 4\pi r^2 dr. \qquad [12]$$

Thus the rate of capture by a single particle is

$$R_{C_1} = \int_{r_p}^{(L+r_p+r_0)} \varrho_{(r)} \cdot P_{(r)} = \int_{r_p}^{(L+r_p+r_0)} \pi R_i (r_p + \bar{K}r)^2 dr. \qquad [13]$$

Prior to integrating equation [13], it is instructive to calculate the value of $[L]$ for our experimental conditions using equations of *Fitch* and *Tsai* (8):

$$L = 3.6 \times 10^{-4} \text{ dm}.$$

Thus the upper limit can be taken as L since $L \gg (r_p + r_0)$ and integration of equation [13] gives

$$R_{C_1} = \pi R_i [L r_p^2 + \bar{K}L(2+\bar{K}) r_p^2 + (2\bar{K}^2 + 3\bar{K} + 1)\bar{K}L^2 r_p + (\bar{K}^2 + \bar{K} + \tfrac{1}{3})\bar{K}^2 L^3]. \qquad [14]$$

Using equation [10], the magnitude of $\bar{K}$ may be obtained from the following values for the experimental constants for the polymerizations which are described in this paper:

$k_1 = 3 \times 10^{-10}$ dm mole$^{\frac{1}{2}}$g$^{-\frac{1}{2}}$ (10)
$M_m = 100$ g mole^{-1}
$k_p = 350$ l mole^{-1}s^{-1} (11)
$[M]_{\text{cont}} = 3.5 \times 10^{-2}$ mole l^{-1}
$D_p = 2 \times 10^{-9}$ dm^2 s^{-1} (see Appendix)
$\bar{D}_0 = 1 \times 10^{-8}$ dm^2 s^{-1} (see Appendix)

from which $\bar{K} = 6.8 \times 10^{-5}$. Therefore equation [14] may approximated to

$$R_{C_1} = \pi R_i (L r_p^2 + \bar{K}L^2 r_p + \tfrac{1}{3}\bar{K}^2 L^3). \qquad [15]$$

The overall rate of capture is obtained by summing the individual rates over all the particles in the system:

$$R_C = \pi R_i (\sum_i L N_i r_{p_i}^2 + \sum_i \bar{K}L^2 N_i r_{p_i} + \tfrac{1}{3}\bar{K}^2 L^3 N). \qquad [16]$$

When equation [15] is compared to equation [1], it is clear that the last two terms reflect the corrections due to oligomer size and the thermal motion of the particles.

The "crowding effect" (assumption (c)) arises when the spheres of action of two or more particles overlap. This is illustrated in Figure 2 in which three particles, A, $B + C$, are each surrounded by a spherical region of radius $L + r_p$ from which they may capture radicals.

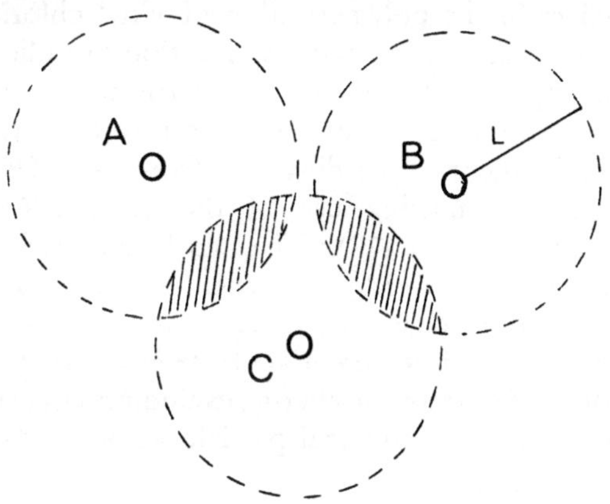

Fig. 2. The "Crowding Effect"

This sphere of action of particle C overlaps those of A and B, so that a radical generated in the cross-hatched zone has a finite probability of being captured by two particles. But the presence of particle C in no way interferes with either the random motion of the radical nor the collision cross-sections, $P_{(r)}$, offered by A or B. The individual capture rates are thus purely additive as given by equation [16]. Should the crowding become so great that there is a significant probability that one particle is interposed between a radical and another particle, the individual capture rates are no longer additive. This error may be estimated from the volume fraction of the system occupied by the particles at the time when the total "capture volume", $\pi r_p^2 L$, becomes unity and $dN/dt = 0$. Under our experimental conditions the polymer occupies less than 1% of the total volume.

Assumption (d), that there are no concentration gradients, is implicit in the collision theory. The diffusion theory, on the other hand, assumes that there *are* gradients in radical concentration in the interfacial region between polymer particles and the continuous medium. This problem is dealt with in the next section.

The final assumption (e) concerns the possibility of an electrostatic interaction between a charged free radical and a charged particle. Such an interaction would be reflected in a dependence of the rate of polymerization on the ionic strength under conditions where the rate of radical capture would affect the overall rate, e. g. in *Smith-Ewart* Case 1 kinetics.

Ugelstad and coworkers found no such salt effect in the polymerization of vinyl chloride where desorption and readsorption of radicals is important (4). We can furthermore estimate the magnitude of the interaction from electrostatic theory according to *Verwey* and *Overbeek* by making use of the *Debye-Hückel* approximation (12). The work required to force a singly charged particle to collide with a multiply charged particle will be equal to $\Psi_0 e$, the surface potential of the larger particle times the electronic charge, excluding viscosity effects. For a colloidal particle whose surface charge, Q, is known,

$$\Psi_0 = \frac{Q}{r_p \varepsilon (1 + \varkappa r_p)}, \qquad [17]$$

where $\varkappa$ is the *Debye-Hückel* constant and ε is the dielectric constant of the medium. *Fitch* and *Tsai* (8) estimated that the charge density of self-stabilized particles prepared in the absence of emulsifier would be on the order of

$$\sigma = 0.65\ \mu\ \text{Coulomb cm}^{-2},$$

a value close to the 0.50 μC cm^{-2} found by *Vanderhoff, van den Hul, Tausk* and *Overbeek* for self-stabilized polystyrene latices (13) and the 0.32 μC cm^{-2} calculated by *Dunn* and *Chong* for polyvinyl acetate latices (14). Assuming a maximum particle radius, r_p, of 4.5 nm (Ref. 8, table 8, p. 89) and an electrolyte concentration of 1×10^{-3} M as used in the experiments described below, equation [17] gives for $\sigma =$ 0.65 μC cm^{-2}

$$e \cdot \Psi_0 = 4.4 \times 10^{-14}\ \text{erg},$$

just equal in magnitude to the thermal energy at 20 °C.

Furthermore, the approaching radical has a growing, hydrophobic tail which upon thermal agitation can "tunnel under" the electrostatic barrier, experiencing *van der Waals* attraction in a manner analogous to the adsorption of a soap molecule at an interface containing previously adsorbed soap molecules. The particles used in the seeded polymerizations described below have radii of from 21.5 to 70 nm. If these have the same surface charge density, they will have surface potentials on the order of 10 kT. The potential drops very rapidly with the distance from the particle, however, as

$$\Psi_r = \frac{Q}{\varepsilon r} \cdot \frac{e^{\varkappa(r_p - r)}}{1 + \varkappa r_p}. \qquad [18]$$

Thus at approximately one half the double layer thickness the potential has dropped to half of its value at the interface[1]), so that the "tunneling effect" becomes highly probable. In conclusion, an electrostatic interaction clearly can be ignored for small particles produced by self-nucleation of oligomeric radicals, but for particles larger than ca. 4.5 nm radius and with surface charge densities much higher than 0.65 μCcm^{-2} and at low ionic strengths, one must proceed with caution. Incidentally, *Fitch, Sprick* and *Prenosil* earlier had asserted that the "energy of activation" resulting from the electrostatic interaction between radical and primary particle should be of the order of 30 kcal per mole or 50 times kT. This seems quite unlikely in the light of these calculations.

Diffusion theory

In emulsion polymerizations involving monomers with negligible chain transfer to

[1]) Calculated for $\varkappa r_p = 4$, which would obtain when $r_p = 40$ nm in 1×10^{-3} molar 1—1 electrolyte.

monomer we may assume that once an oligomeric radical enters a polymer particle, it stays there. The *van der Waals* forces within the particle are strong and the chain radical quickly grows in the environment of relatively high monomer concentration. This means that the concentration of oligomer radicals in the continuous medium drops to zero at the polymer interface, but has some finite value, C_s, in the medium far from the interface. In the immediate environment of the particle, there will then be a concentration gradient, creating a diffusion flux of radicals towards the particle. This situation is shown schematically in fig. 3 where r_h is the distance from the center of the polymer particle (solid circle) to the edge of this "diffusion layer" (dashed circle), and C_0 is the oligomeric radical concentration.

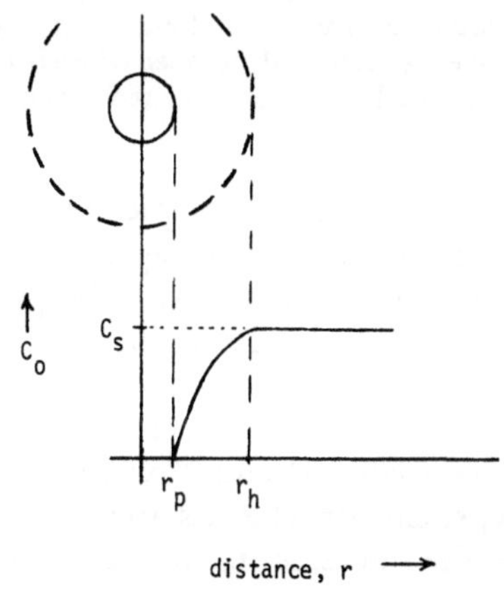

Fig. 3. The Diffusion Layer of Oligomeric Radicals Surrounding a Polymer Particle (Solid Circle)

Integration of *Fick*'s First Law equation with the boundary conditions:

$$C_0 = 0 \text{ at } r = r_p$$

$$C_0 = C_s \text{ at } r = r_h,$$

gives the well known expression for spherical particles for the diffusion flux, J, of radicals to a particle:

$$J = \frac{4\pi \bar{D}_{0p} C_s r_p r_h}{r_h - r_p}. \qquad [19]$$

The flux is, of course, equal to the rate of capture: $J = R_{C_1}$. The thickness of the diffusion boundary layer is not known, but when the particles are small it may be safe to assume that $r_h \gg r_p$ (5, 6).

Adding the individual fluxes for all the particles then gives the overall rate of capture:

$$R_c = \sum_i J_i = 4\pi \bar{D}_{0_p} C_s \sum_i N_i r_{p_i}, \qquad [20]$$

which for monodisperse colloids becomes simply

$$R_c = 4\pi \bar{D}_{0p} C_s N r_p.$$

The rate of capture in this case is proportional to the particle radius, whereas in the simplified collision theory it is proportional to the square of the radius (equation [1]).

Experimental

Two sets of experiments were performed, Series I in which seeded polymerizations were run, and Series II in which homogeneous conditions obtained at the start.

Series I

Four seed latices were prepared by methods previously described (8). They were polymethyl methacrylate latices that were stabilized solely by sulfate or sulfonate end groups on the polymer chains introduced by initiator. No emulsifier was used. The size distribution was determined by electron microscopy, and the number-, surface- and weight-average diameters, $\bar{d}_n$, $\bar{d}_s$ and $\bar{d}_w$, respectively, are given in table 1.

Table 1. Particle Sizes of Seed Latices. All sizes × 10^7 dm

Seed Latex No.	$\bar{d}_n$	$\bar{d}_s$	$\bar{d}_w \times 10^7$ dm
1	11.5	14.0	14.1
2	10.2	12.0	12.5
3	7.0	8.6	9.0
4	3.3	4.3	4.8

The size distributions of these latices were generally characterized by a rather narrow principal peak along with a broader distribution of fewer smaller particles. These are shown in figs. 4 and 5.

The purpose in Series I was to determine at what critical concentration of seed latex no new particles would be formed in seeded polymerizations. Both the size and concentration of seed particles were varied. This critical concentration could then be compared to that predicted by theory. It was therefore necessary to ensure that each polymerization was run at the

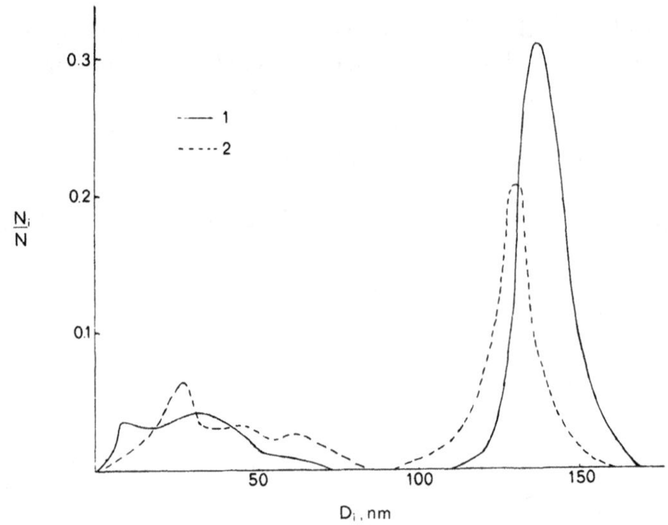

Fig. 4. Particle size distribution of PMMA Seed latexes 1 and 2

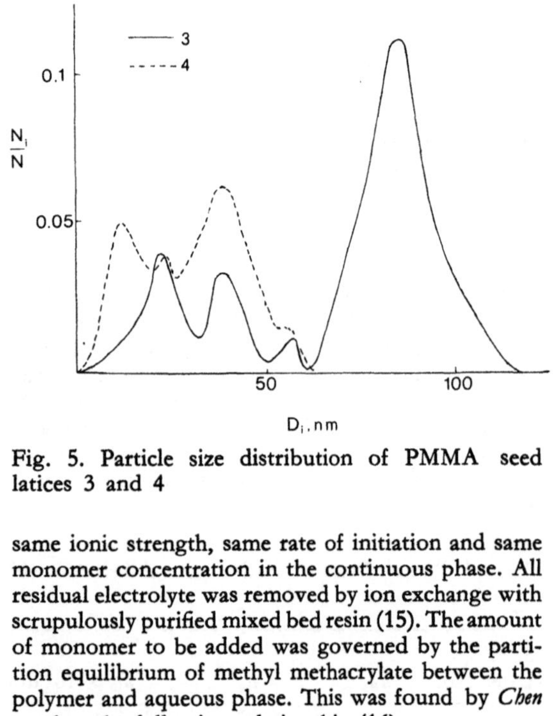

Fig. 5. Particle size distribution of PMMA seed latices 3 and 4

same ionic strength, same rate of initiation and same monomer concentration in the continuous phase. All residual electrolyte was removed by ion exchange with scrupulously purified mixed bed resin (15). The amount of monomer to be added was governed by the partition equilibrium of methyl methacrylate between the polymer and aqueous phase. This was found by *Chen* to obey the following relationship (16):

$$[M]_p = 917\,[M]_{aq}^{1.88}\,.$$

All seeded runs were made with a monomer concentration of 3.5×10^{-2} mole l^{-1} in the aqueous phase.

The initiator system was the same in all cases:

$[K_2S_2O_8] = 8.46 \times 10^{-4}$ mole l^{-1}

$[NaHSO_3] = 1.30 \times 10^{-3}$ mole l^{-1}

$[FeSO_4 \cdot (NH_4)_2SO_4 \cdot SO_4 \cdot 6H_2O]$
$= 2.74 \times 10^{-6}$ mole l^{-1}

which was calculated (17) to give a rate of radical production of $7.74. \times 10^{16}\ l^{-1}s^{-1}$ at 30 °C. Precautions taken to ensure purity of reagents, water and nitrogen were the same as those described earlier (8). The stability of the systems towards flocculation by the initiator salts was checked by light scattering. Concentrations of 1–1 electrolyte over 100 times those used in the polymerizations were required to produce any measurable change (18).

The results of fifteen seeded polymerizations are summarized in Table 2, along with the original surface-average size and number of each of the four seed latices used and the calculated change in particle number, ΔN.

Figs 6 and 7 give examples of the primary data from which the averages in table 2 are calculated. In fig. 6 is shown the particle size distribution of the latex formed from seed latex No. 4 in Exp. No. 4–1. The peaks labelled 1 and 2 in the seed have grown to those designated 1′ and 2′ while a generation of new, small particles has appeared at peak 0. Hence the increase in number, $\Delta N = 4.82 \times 10^{14}$ particles per liter. The major peak has become narrower, and therefore higher, typical of seeded latex polymerizations (19). In Exp. No. 3–5, on the other hand, the number and size of seed particles were sufficient initially to capture all the radicals produced so that almost no new particles were formed. This is shown in fig. 7 where essentially only growth can be observed.

The initial overall rates of capture were computed for each of the experiments in Series I by use of equation [16]. It turned out that the last two terms in that equation, representing the corrections for oligomer size and particle motion together were never larger than 1.4% of the first term. They were therefore not employed, so that the simple collision equation [1] was used and the capture

Table 2. Results of seeded polymerizations: change in the average number of particles. N_s and N_f are seed and final particle concentrations, respectively

Seed Latex	Exp. No.	$N_s \times 10^{15}\ 1^{-1}$	$N_f \times 10^{15}\ 1^{-1}$	$\Delta N \times 10^{15}\ 1^{-1}$
1. $\bar{d}_s$ = 1400 Å	1–2	3.39	6.85	3.46
	1–3	4.75	5.29	0.54
	1–4	5.15	6.71	1.56
	1–5	5.42	5.24	−0.18
2. $\bar{d}_s$ = 1200 Å	2–1	1.72	3.16	1.44
	2–2	2.29	3.77	1.48
	2–3	3.43	9.84	6.41
	2–4	4.58	5.17	0.59
3. $\bar{d}_s$ = 860 Å	3–1	2.69	4.34	1.65
	3–2	3.59	5.51	1.92
	3–3	5.38	7.09	1.71
	3–5	8.72	9.14	0.42
4. $\bar{d}_s$ = 430 Å	4–1	9.78	14.6	4.82
	4–2	13.0	15.6	2.60
	4–3	19.6	19.8	0.20

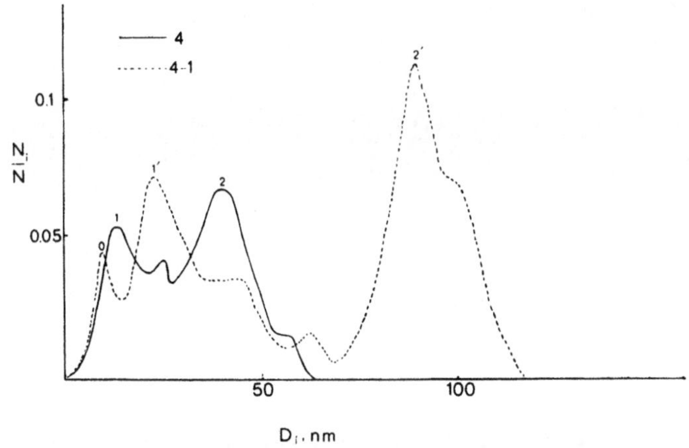

Fig. 6. Particle size distribution from Exp. No. 4–1 compared to that of original seed latex No. 4

rates calculated from the entire particle size distribution in each experiment by summation. The values of R_C so obtained were divided by the experimental R_t to give a reduced rate of capture, which should approach unity as ΔN approaches zero. This is based upon the kinetic theory for particle formation proposed by *Fitch* and *Tsai* (8) in which

$$\frac{dN}{dt} = R_t - R_C, \qquad [21]$$

so that when $dN/dt \to 0$, $R_C \to R_t$, or $R_C/R_t \to 1$. The calculated reduced rates of capture are given in table 3. Finally the experimental

Fig. 7. Particle size distribution from Exp. No. 3–5 compared to that of original seed latex No. 3

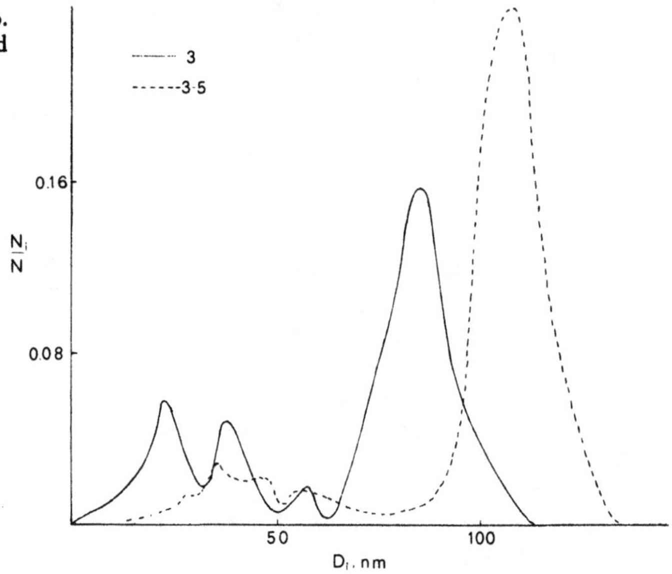

Table 3. Computed values for the reduced rate of capture, R_c/R_i

Exp. No.	R_c/R_i	Exp. No.	R_c/R_i	Exp. No.	R_c/R_i	Exp. No.	R_c/R_i
1–2	1.49	2–1	0.58	3–1	0.44	4–1	0.36
1–3	2.08	2–2	0.80	3–2	0.58	4–2	0.46
1–4	2.25	2–3	1.16	3–3	0.68	4–3	0.71
1–5	2.37	2–4	1.54	3–5	1.40		

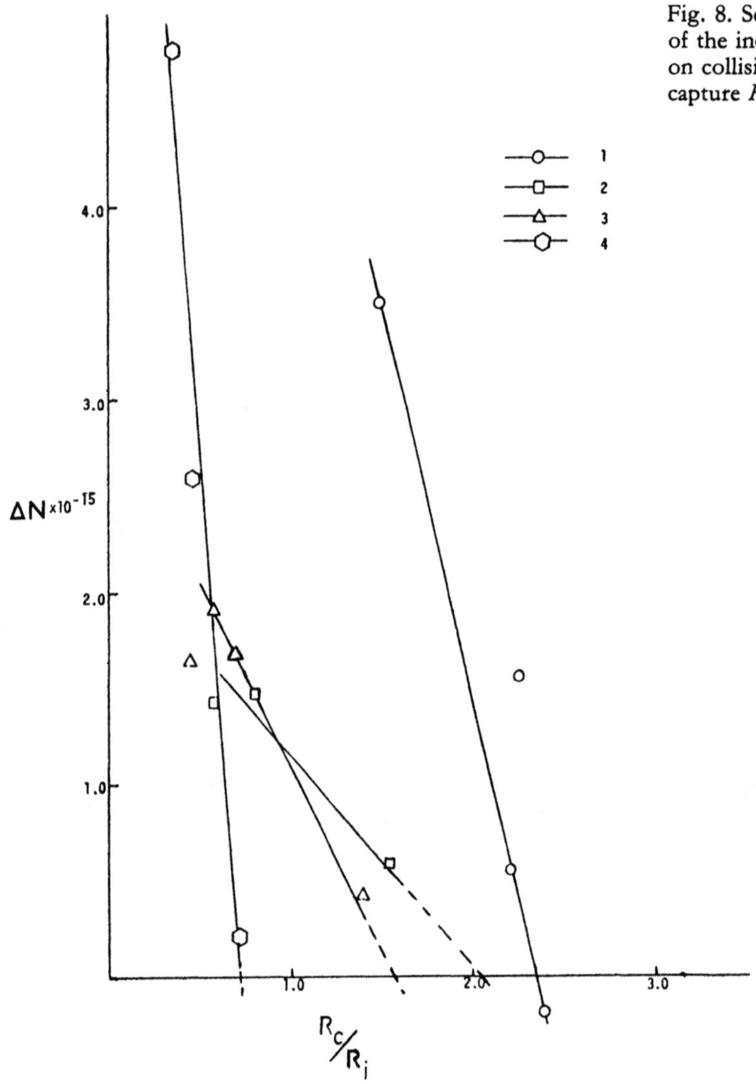

Fig. 8. Seeded polymerizations: dependence of the increase in number of particles, ΔN, on collision theory values of reduced rate of capture R_c/R_i

values of ΔN (table 2) were plotted against the computed values of R_c/R_i (table 3) and extrapolated or interpolated to $\Delta N=0$ as shown in fig. 8. The values of the intercepts correspond to the concentration of seed latex which is just sufficient to capture all the free radicals produced in the aqueous phase as predicted by the collision theory. Table 4 gives these intercepts as taken from fig. 8. The predicted values are all of the

Table 4. Calculated intercept values. The two theories compared

Seed Latex Series	$\left(\frac{R_c}{R_i}\right)_{\Delta N=0}$	$\frac{293}{r_p}\cdot\left(\frac{R_c}{R_i}\right)_{\Delta N=0}$
1	2.35	0.98
2	2.05	1.00
3	1.55	1.05
4	0.73	0.99
Avge.	1.67	1.01

correct order of magnitude with an average of 1.67 and a total variation within a factor 3. The collision theory values, which depend on the particle radius squared may be converted by an empirical factor of $293/r_p$ to reduced capture rates proportional to the radius of the seed particles as required by the diffusion theory. In this case the agreement is excellent. We have made the tacit assumption that the (presumably steady state) concentration of radicals, C_s, is the same for all experiments in order to convert capture rates calculated by equation [16] to those calculated by equation [20].

Series II.

In Series II experiments, the initial system was homogeneous. Under conditions where no flocculation occurs, the kinetics of particle formation are described by equation [21]. When the steady state, $dN/dt = 0$, is just achieved,

$$\text{Steady state: } R_i = R_C, \quad [22]$$

and the collision theory for monodisperse colloids gives:

$$\text{Steady state: } R_C = R_i \pi r_p^2 NL = R_i, \quad [23]$$

$$\text{or } Nr_p^2 = \frac{1}{\pi L}. \quad [24]$$

The radical diffusion distance, L, is a function of the initial monomer concentration: (8)

$$L = K_3/[M]_0^{\frac{1}{2}}, \quad [25]$$

where K_3 is a constant.

Combining equations [24] and [25] gives

$$\frac{Nr_p^2}{[M]_0^{\frac{1}{2}}} = K_4, \quad [26]$$

where K_4 is a constant according to the collision theory.

This series of emulsion polymerizations of methyl methacrylate was carried out with varying amounts of the emulsifier, sodium dodecyl sulfate (*SDS*) and monomer (*MMA*). A constant rate of initiation was employed throughout, the same as in Series I. The temperature was again 30 °C. The monomer concentrations were all below saturation and the emulsifier below the critical micelle concentration, so that the initial solutions were homogeneous. The techniques for polymerization and electron microscopic particle size measurements have already been described (8). The amount of emulsifier added was sufficient to prevent flocculation and to produce essentially monodisperse colloids. The particle sizes and number concentrations are given in table 5. From these, the products Nr_p^2 were calculated and are given in table 5 also. Because the monomer concentration was varied, it was necessary to divide the products by $[M]_0^{\frac{1}{2}}$ since the parameter L in equation [16] is inversely proportional to the square root of the monomer concentration (8). The radii

Table 5. Comparison of the two theories. Final particle sizes and numbers for initially homogeneous emulsion polymerization

$[M]_0 \times 10^2$ mole l^{-1}	$SDS \times 10^4$ mole l^{-1}	$N \times 10^{-16}$ l^{-1}	$r_p \times 10^9$ dm	$\frac{Nr_p^2}{[M]_0^{\frac{1}{2}}} \times 10^{-4}$ $(\text{mole}^{-1}\text{dm})^{\frac{1}{2}}$	$\frac{Nr_p}{[M]_0^{\frac{1}{2}}} \times 10^{-11}$ $(\text{mole}\cdot\text{dm})^{-\frac{1}{2}}$
0.380	27.8	48.8	55	2.39	4.35
1.90	27.8	56.0	90	3.29	3.65
3.50	11.1	64.5	103	3.66	3.55
	27.8	63.8	104	3.69	3.55
5.71	13.9	87.2	112	4.58	4.09
	27.8	75.0	116	4.22	3.64
7.61	13.9	66.4	135	4.39	3.25
	27.8	80.0	126	4.60	3.65
9.51	13.9	81.3	136	4.88	3.59
	17.4	102	126	5.25	4.17
	20.8	131	116	5.72	4.93
	27.8	90.5	131	5.04	3.84

given in table 5 are the final values after 100% conversion, rather than those taken at the critical moment when dN/dt becomes zero. The radius at this critical conversion, C_c, would be obtained by multiplying the final radius by the factor $(C_c/100)^{1/3}$, where C_c is likely to vary only slightly with $[M]_0$. Since $(C_c)^{1/3}$ is unknown, but a very weak function of $[M]_0$, we have ignored this correction. Again as in Series I, the values $Nr_p/[M]_0^{1/2}$ reflect the predictions of diffusion theory only if C_s is constant for all experiments.

Conclusions

The experimental evidence from the seeded and initially homogeneous emulsion polymerizations of methyl methacrylate show that the rate of capture of the growing oligomeric radicals in the continuous phase by the polymer particles is proportional to $N \cdot r_p$. This also supports the contention that the capture process is diffusion-controlled. The matter cannot be considered as settled since the dependence of the concentration of radicals, C_s, on the experimental variables has not yet been deduced. Furthermore, the electrostatic interaction, although thought to be small, may still play a role. More experiments with uncharged radicals are needed.

These findings are important for use in predicting particle formation kinetics in polymer colloids, determining correct initial conditions for seeded polymerizations and true emulsion polymerization in which initiation occurs within emulsified monomer droplets, calculating particle growth kinetics in *Smith–Ewart* Case 1 polymerizations and those in which desorption and readsorption of free radicals is involved.

Appendix

Estimation of diffusion coefficients D_p and $\bar{D}_0$

To obtain D_p the *Stokes–Einstein* diffusion equation for hard spheres may be used:

$$D = \frac{kT}{6\pi\eta r_p} \,.$$

The average radii of the seed particles used in Series I ranged from 2×10^{-7} dm to 7×10^{-7} dm. These will then have values of D_p from 3×10^{-10} to 1×10^{-9} $dm^2 s^{-1}$.

The value of $\bar{D}_0$ is the time-average quantity during the period of growth of the oligomeric radical in solution. There are apparently no published values of D_0 for polymethyl methacrylate of low molecular weight in aqueous solution. A good estimate can be made from diffusion coefficients for this polymer in other solvents (21) and for some natural polymers of comparable molecular weights and radii of gyration in aqueous solutions (22):

$$\bar{D}_0 = 1 \times 10^{-8}\ dm^2\ s^{-1}.$$

List of Symbols

Symbol	Description	Dimensions
R_{C_1}	Rate of capture of radicals by single particle	s^{-1}
R_i	Rate of generation of effective radicals	$l^{-1}s^{-1}$
r_p	Radius of a polymer particle	dm
r	Radial distance from the center of a particle	dm
L	Distance an oligomeric radical diffuses before self-nucleation	dm
D_p	Diffusion coefficient of a polymer particle	dm^2s^{-1}
r_0	Radius of gyration of an oligomeric free radical	dm
A	Collision cross-section	dm^2
k_1	Proportionality constant between molecular weight and radius of gyration	$dm\ g^{-1/2} mole^{1/2}$
M_0	Molecular weight of an oligomeric radical	$g\ mole^{-1}$
X	Degree of polymerization	—
M_m	Molecular weight of a monomer unit	$g\ mole^{-1}$
$[M]_{cont}$	Concentration of monomer in in continuous phase	$mole\ l^{-1}$
k_p	Specific rate constant for chain propagation	$l\ mole^{-1}s^{-1}$
k_2	A constant relating oligomer size to time of growth	$dm\ s^{-1/2}$
$\bar{D}_0$	Diffusion coefficient of oligomer radical averaged over its time of growth	$dm^2\ s^{-1}$
$P_{(r)}$	Probability of collision	—
$\bar{K}$	Proportionality constant between radius of gyration of a radical and distance travelled	—
$\varrho_{(r)}$	Differential rate of effective radical generation	s^{-1}
N_i	Number concentration of particles of radius r_{p_i}	l^{-1}
N	Total number concentration of particles	l^{-1}
Ψ_0	Surface electrical potential	V
e	Electronic charge	C
ε	Dielectric constant	—
Q	Electrostatic charge	C
$\varkappa$	*Debye–Hückel* reciprocal double layer thickness	dm^{-1}
C_0	Number concentration of oligomeric radicals	l^{-1}

C_s	Number concentration of oligomeric radicals in the continuous medium far from a particle	l^{-1}
r_h	Radius of diffusion boundary layer	dm
$\bar{D}_{op}$	Average diffusion coefficient for relative motions of polymer particle and oligomer radical	$dm^2\ s^{-1}$
J	Diffusion flux	$dm^{-2}s^{-1}$
$\bar{d}_n, \bar{d}_s, \bar{d}_w$	Number-, surface- and weight-average particle diameters	dm
$[M]_0$	Initial monomer concentration	mole l^{-1}
K_3	Constant of proportionality between L and $[M]_0^{1/2}$	$mole^{-1/2}dm^{5/2}$
K_4	Collision constant	$mole^{-1}dm^{1/2}$
η	Viscosity of fluid medium	$Ns\ dm^{-2}$
k	*Boltzmann* constant	$J\ deg^{-1}$
T	Absolute temperature	deg
t	Time	s
	NB: 1 and dm^3 are used interchangeably.	

Summary

In emulsion polymerization free radicals generated in the continuous medium enter the monomer/polymer particles at a rate which is governed by either collision or diffusion dynamics. Previously various theories have assumed one or the other model. Two series of experiments are described to test which theory applies. One series involves seeded emulsion polymerization, while the other starts with a perfectly homogeneous solution. The rate of radical "capture" by the monomer/polymer particles in both series is proportional to the product of particle number and radius, in support of a diffusion-controlled process. The electrostatic interaction between like charged radicals and particles is thought to be negligible, but experimental proof is required.

Zusammenfassung

Bei der Emulsionspolymerisation greifen die in dem kontinuierlichen Medium erzeugten freien Radikale die monomeren/polymeren Partikel in einer Geschwindigkeit an, die entweder von Kollisions- oder Diffusionsdynamik beherrscht wird. Früher haben verschiedene Theorien das eine oder das andere Modell anerkannt. Zwei Serien von Experimenten werden beschrieben, die testen sollen, welche Theorie gültig ist. Bei der einen Serie wird von der Emulsionspolymerisation mit Keimen (seeded emulsion polymerization) ausgegangen, während die zweite mit einer völlig homogenen Lösung startet. Die Geschwindigkeit der ‚Einfangung' von Radikalen durch die monomeren/polymeren Partikel ist in beiden Serien proportional zum Produkt aus Partikelzahl und Radius, was einen Prozeß unterstützt, der von Diffusion beherrscht wird. Die elektrostatische Wechselwirkung zwischen Radikalen und Partikeln gleicher Ladung wird als vernachlässigbar betrachtet, doch ist der experimentelle Nachweis dafür noch zu erbringen.

References

1) *Smith, W. V.*, and *R. H. Ewart*, J. Chem. Phys. **16**, (6), 592 (1948).

2) *Smith, W. V.*, J. Amer. Chem. Soc. **70**, 3700 (1948).

3) *Ugelstad, J., H. Fløgstad, F. K. Hansen*, and *T. Ellingsen*, J. Polym. Sci. Symposium No. 42, 473 (1973).

4) *Ugelstad, J., P. C. Mørk, P. Dahl*, and *P. Ranges*, J. Polym. Sci. Part C. No. 27, 49 (1969).

5) *Reiss, H.*, and *V. K. LaMer*, J. Chem. Phys. **18**, 1 (1950).

6) *Reiss, H.*, J. Chem. Phys. **19**, 482 (1951).

7) *Gardon, J. L.*, J. Polym. Sci. A–1, **6**, 623 (1968).

8) *Fitch, R. M.*, and *C. H. Tsai*, in: *R. M. Fitch*, (Ed.), Polymer Colloids, pp. 73–116, New York 1971.

9) *Alexander, A. E.*, and *D. H. Napper*, Vol. 3, *A. D. Jenkins*, (Ed.), Progress in Polymer Science, pp. 145–197. (New York 1971).

10) *Tanford, C.*, Physical Chemistry of Macromolecules, (New York 1961).

11) *Matheson, M. S.*, et al., in *C. Walling*, Free Radicals in Solution, p. 94, (New York, 1957).

12) *Verwey, E. J. W.*, and *J. Th. G. Overbeek*, Theory of the Stability of Lyophobic Colloids, p. 37, (New York 1948).

13) *Vanderhoff, J. W., H. J. van den Hul, R. J. M. Tausk*, and *J. Th. G. Overbeek*, Proceedings of Symposium on Clean Surfaces, Raleigh, N. C. April, 1968.

14) *Dunn, A. S.*, and *L. C. H. Chong*, Brit. Polym. J. **2**, 49 (1970).

15) *van den Hul, H. J.*, and *J. W. Vanderhoff*, in: *R. M. Fitch* (Ed.), Polymer Colloids, p. 3, (New York 1971).

16) *Chen, T. J.*, Ph. D. Thesis, North Dakota State University, Fargo (1967).

17) *Fritzsche, P.*, and *J. Ulbricht*, Faserforsch. Text. Techn. **14**, 517 (1963).

18) *Fitch, R. M.*, and *L. B. Shih*, unpublished results.

19) *van den Hul, H. J.*, and *J. W. Vanderhoff*, Brit. Polym. J. **2**, 121 (1970).

20) *Fitch, R. M.*, Brit. Polym. J., **5**, 467 (1973).

21) *Flory, P. J.*, Principles of Polymer Chemistry, (Ithaca, N. Y. 1953).

22) *Brandrup, J.*, and *E. H. Immergut*, (Ed's.), Polymer Handbook, (New York 1967).

Authors' address:

R. M. Fitch and *Lih-bin Shih*
Institute of Materials Science
University of Connecticut
Storrs, Conn. 06268 (USA)

Progr. Colloid & Polymer Sci. **56**, 12–15 (1975)

2

Royal Institute of Technology, Stockholm 70 (Sweden)

Liquid Water at Interfaces

E. Forslind

With 2 figures and 2 tables

(Received January 17, 1974)

I shall make use of the interstitial lattice model of the water structure (1) to discuss some phenomena at the water solid interface. This structure model is based on three principal ideas:

i) Liquid water shows strong affinities with crystalline matter (*Bernal* and *Fowler*, 1933) (2).
ii) The crystalline lattice of ice is molecular in spite of intermolecular proton transfer (*Pauling*, 1935) (3).
iii) The structure of liquid water may be derived from that of ice Ih (*Samoilov* (4), 1946; *Forslind* (5), 1952; *Danford* and *Levy* (6), 1962).

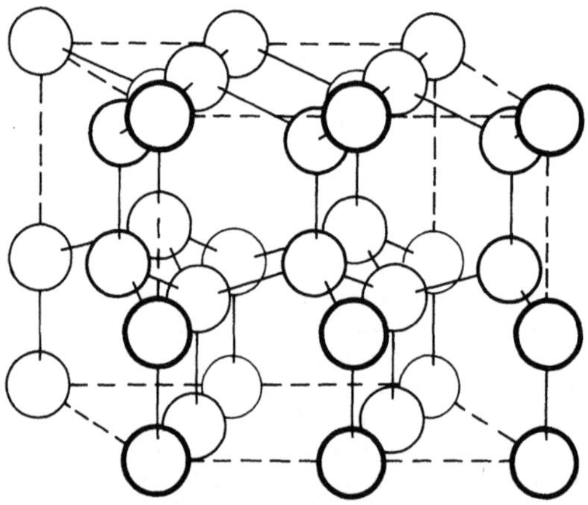

Fig. 1. The ideal lattice of ice Ih. Only the oxygen atoms are shown (*Forslind*, 1952) (5)

The cell dimensions of the ideal ice Ih lattice (7, 8) (fig. 1) are $a = 4.5226$ Å, $c = 7.3670$ Å at 0 °C. The hexagonal unit cell of the oxygen lattice contains four molecules H_2O in approximately tetrahedral coordination only, since the axes ratio $c/a = 1.62893$ deviates from the theoretical value $c/a = \sqrt{\frac{8}{3}} = 1.63299$ of the ideal tetrahedral structure. Since the deviation is larger than the reported error of observation, it is interesting to note that *Bjerrum* in a discussion of the structure of ice, suggested that the distance between atoms in neighbouring basal layers, possessing mirror symmetry, should be slightly shorter than the bonds between atoms in central symmetric arrangement. This would imply that the hydrogen bond along the hexagonal c-axis is weaker, which is in good agreement with experimental data regarding the plastic deformation of ice single crystals.

The ideal ice Ih oxygen lattice, shown in fig. 1, should then be modified by the introduction of slightly bent and, consequently, weaker bonds in the direction of the c-axis. In such a modified model of the ice lattice, changes in hybridization of the water molecules may be expected to occur, implying the appearance of non-linear restoring forces and coupling between valence and bending vibrations. With rising temperature, the increasing anharmonicity of the intermolecular vibration will lead to fluctuations in the distribution of the thermal amplitudes, producing localized amplitude peaks that eventually lead to disrupture of hydrogen bonds. If the thermal amplitude is sufficient, the molecule that is made free will pass through the face of the surrounding coordination tetrahedron to take up an interstitial position in the lattice. The arising lattice defect, a so-called *Frenkel* defect, consists of one vacant lattice point and one interstitial, non-associated water molecule.

Just below and at the melting-point, a certain number of *Frenkel* defects will appear in the ice lattice[1]). If we designate the negative parts of the *Frenkel* defects, the vacant lattice points, as "holes", we find that the holes and the interstitial molecules may be dissociated, and each part of the defect will be free to diffuse in the lattice. For this diffusion, however, a certain activation energy is required and, when provided, a number of holes will eventually reach the boundaries of the system, where they will be annihilated and leave behind lone interstitial molecules in the lattice. This annihilation of holes will produce a volume decrease and a corresponding increase in density due to the lone interstitial molecules. As long as both components of the *Frenkel* defects are equal in number, the volume change produced by the defects can be neglected in comparison to the volume changes just described, due to the annihilation of the vacant lattice sites. The annihilation of holes at the boundaries of the system corresponds to the melting process and the latent heat of fusion of ice is essentially determined by the energy of formation of the lone interstitial molecules, which we henceforth shall call *Schottky* defects.

For an approximate thermodynamical treatment of the water lattice it is now simply a question of establishing the conditions of equilibrium between the molecules in lattice positions, the holes and the interstitial molecules, as functions of the temperature.

Performing this calculation (5), we find that the number of *Schottky* defects in water, according to our structural model, may be derived from the volume change in the ice/water transition at 0 °C and the latent heat of fusion. The fraction of *Schottky* defects then amounts to about

$$m_s = 0.077$$

while the fraction of *Frenkel* defects obtains a value of

$$m_F = 0.084,$$

the total defect fraction of all the water molecules present being

$$m_s + m_F = 0.161 = m,$$

which is also equivalent to the total fraction of interstitial molecules.

Since the fraction of *Frenkel* defects corresponds to the fraction of holes, i.e. vacant lattice sites, and since each vacancy corresponds to four three-coordinated molecules, the fraction of three-coordinated molecules amounts to about 0.336 of the total number of the molecules present. The lattice molecules comprise 0.839 of the total number. Hence about 0.503 of the total number of molecules present are four-coordinated, assuming with *Walrafen* (9) that no other types of coordination occur at 0 °C. The computed lattice defect and coordination types may then be summarized in table 1.

Table 1

Percentage of molecules in water at 0 °C of coordination number:

4	3	2	0
50.3	33.6	0	16.1

The prevalence of three- and four-coordinated lattice molecules is maintained up to +25 °C below which temperature, according to *Walrafen* (9, 10), no two-coordinated molecules appear in the water lattice. Combining the experimental *Raman* intensity data given by *Walrafen* (10) and the temperature dependence of the total fraction of interstitial water molecules computed by *Forslind* (11), the temperature dependence of the various coordination species between 0 °C and 100 °C may be estimated (12) as shown in fig. 2, in which the index, *i*, on the fraction *f*, indicates the coordination number.

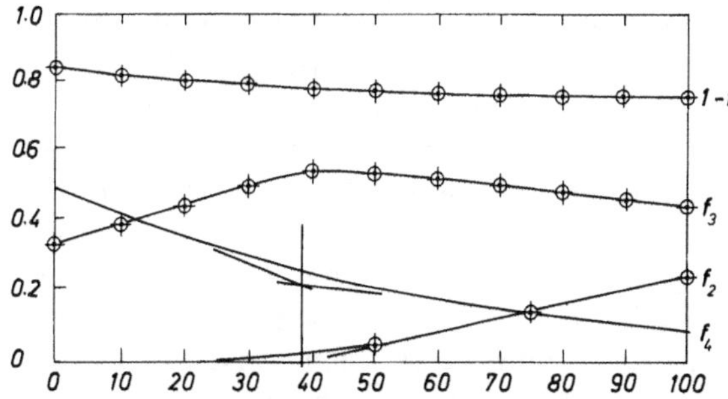

Fig. 2. Temperature dependence of the various coordination species in water computed on the basis of *Walrafen's* (10) *Raman* intensity data (*Forslind*) (11). End point tangents to the f_4 curve cross at 38 °C

[1]) In this discussion of principles, the arguments refer to the single crystal lattice, because in the polycrystalline case other types of lattice defects will be of giant influence.

It is noted that a break in the rise of the f_3 curve occurs at around 38 °C coincident with a significant rise in the number of two-coordinated water molecules that started at about 25 °C.

The maximum in the three-coordinated species appearing at about 40 °C represents in fact a point of instability of the water lattice vibrations. The progressing expansion of the hydrogen bonds and the increasing anharmonicity effects give rise to strong thermal fluctuations that affect the balance between the four- and three-coordinated species. It appears that a type of umbrella-inversion overtakes the vibration of the three-coordinated molecules at the temperature in question, suddenly increasing the number of two-coordinated molecules and giving rise to a series of events characteristic of a great number of water containing systems. Thus the tropocollagen molecule that is stabilized by water bridges in primary adsorption (*Lindner* and *Forslind* (13), 1973) is denatured at 38 °C. The conformation of the whole polypeptide chain undergoes a change at this temperature due to the breakdown of water bridges that take over the stabilizing effect of the pyrrolidine ring closure occurring in the non-polar part of the chain. Similarly, the cellulose shows a number of characteristic transformations at the same temperature (*Martin-Löf* and *Söremark* (14), 1969; *Forslind* (15), 1971) depending on water bridge stabilization of the polysaccharide chain conformation in the so-called amorphous zones. In these zones of the cellulose I and II lattices, the chain motion is critically determined by the presence of water molecules, so that the water content determines the permeability to diffusion along the chains. There is, in other words, a critical temperature limit of 38 °C above which a cellulose membrane cannot normally retain water or prevent the loss of water from living tissue.

As is well known water is an important component of biomembranes based on the lecithin-cholesterol bi-layer structure. Thus it appears possible to design a membrane model based on the lecithin-water system that displays a number of characteristic properties of the real, complex membrane.

Taking the water lattice symmetry as our point of departure we have designed a lecithin-water complex that fits to the normal water structure and in which the lecithin molecules are linked by water bridges giving rise to a spacing that closely corresponds to the observed molecular cross-section as revealed by X-ray investigations (*Forslind* and *Kjellander*, 1974) (16).

A number of predictions, among other things the thermal instability of the interlamellar water layers, have been derived from the model, which satisfy the experimental conditions available, while others remain to be investigated. Similar structure models containing up to one cholesterol molecule per lecithin have been derived, showing similar correspondence to the observed properties of the real membranes.

I should finally like to exemplify the structuring of water in an inorganic colloidal and lamellar system. In 1954 (17) *Norrish* published a study of the lattice expansion of a sodium montmorillonite as a function of the water content. He observed a series of hydration levels above 20 Å basal spacing, which are compared in table 2 with theoretical data derived from the structure of the adsorbed water. The first three column contains the data derived from stable water conformations, with the exception of the first, which is unstable, and represents the initial stage of hydration. The fourth column contains the observed data. It should be emphasized that the observation of the higher hydration stages is extremely difficult and does not permit the correct assessment of the water content in the sample. The system is as a rule polymorphous and several crystalline aggregates appear simultaneously. It is then extremely satisfactory that the agreement between theory and experiment is as good as table 2 shows. It proves beyond any doubt the structural properties of water in specific adsorption on a crystalline substrate.

Table 2

Number of layers	Theoretical water content g H_2O/g clay	Calculated spacing Å	Observed spacing Å
3	0.179	18.73	18.7
14	0.833	39.89	40.0
16	0.952	43.57	43.8
20	1.190	50.93	49.2
28	1.665	65.65	65.2
32	1.903	73.01	72.0

Summary

Starting from the structural properties of liquid water and making use of the known bond lengths and coordination types of the hydrogen bonded water molecule, a number of interface phenomena characteristic of systems containing water bridges are derived and discussed.

Zusammenfassung

Von den Struktureigenschaften des flüssigen Wassers ausgehend und mit Hilfe der bekannten Längen und Orientierungen der Wasserstoffbrücken des Wassermoleküls wird eine Anzahl von Grenzflächeneigenschaften hergeleitet und besprochen, die charakteristisch für Wasserbrücken enthaltende Systeme sind.

References

1) *Frank, H. S.*, Science, N. Y. **169**, 635 (1970).
2) *Bernal, J. D.* and *R. H. Fowler*, J. Chem. Phys. **1**, 515 (1933).
3) *Pauling, L.*, J. Chem. Soc. **57**, 2680 (1935).
4) *Samoilov, O. Ya.*, Zh. Fiz. Khim. **20**, 1411 (1946).
5) *Forslind, E.*, Acta Polytech. Scand. **115**, 9 (1952).
6) *Danford, M. D.* and *H. A. Levy*, J. Amer. Chem. Soc. **84**, 3965 (1962).
7) *Megaw, H. D.*, Nature, Lond. **134**, 900 (1934).
8) *Owston, P. G.* and *K. Lonsdale*, J. Glaciol. **1**, 118 (1948).
9) *Walrafen, G. E.*, Structure of water. In: *A. K. Carrington* and *P. Jones* (Ed.) Hydrogen-Bonded Solvent Systems, p. 9, (London 1968).
10) *Walrafen, G. E.*, J. Chem. Phys. **44**, 1546 (1966).
11) *Forslind, E.*, Water association and hydrogels. Proc. Second Int. Congr. on Rheology, Oxford, p. 50. Scientific Publications (London 1953).
12) *Forslind, E.*, Quart. Rev. Biophysics **4**, 4, 325 (1971).
13) *Lindner, P.* and *E. Forslind*, Chem. Scripta, **3**, 57 (1973).
14) *Martin-Löf, S.* and *Chr. Söremark*, A phase transformation associated with the loss of water in crystalline α—ω-glucose monohydrate at 37 °C and some aspects on secondary thermal transitions in carbohydrates. Swedish Forest Products Res. Lab. (Stockholm 1969).
15) *Forslind, E.*, NMR Wide Line Studies of Water Sorption and Hydrogen Bonding in Cellulose. NMR-Basic Principles and Progr. Vol. **4**, (Berlin–Heidelberg –Berlin 1971).
16) *Forslind, E.* and *R. Kjellander*, Svensk Naturvetenskap, Stockholm, 1974. (Stockholm 1974).
17) *Norrish, K.*, Nature, **173**, 254 (1954).

Author's address:

E. Forslind
Royal Institute of Technology
Stockholm

Progr. Colloid & Polymer Sci. **56**, 16–20 (1975)

3

Swedish Institute for Surface Chemistry, Stockholm (Sweden)

Microemulsions and solubilization by nonionic surfactants

S. Friberg and I. Lapczynska)*

With 6 figures

(Received January 17, 1974)

Microemulsion, transparent stable "emulsions" of water and hydrocarbons stabilized by a soap and an alcohol of medium chain length have caused a long discussion with regard to their structure. Originally *Schulman* and his collaborators (1–5) favoured the view that these "emulsions" were true emulsions of large droplets. The systems were, however, thermodynamically stable. Investigations by *Shinoda* (6–10) of solubilized systems containing nonionics and of solubilized systems containing a mixture of soap and an alcohol of medium chain length (11) showed that typical microemulsion compositions according to *Schulman* could be obtained by continuous addition of a soap solution in water to an alcohol solution in benzene without any phase transition taking place during the addition. In view of the above, it was difficult to see the advantage of considering the microemulsion as an emulsion, since it required the introduction of a phase transition which could not be observed.

Studies by *Shinoda* and co-workers (6–10) of the temperature dependence of the solubilization capacity of nonionic surfactants gave results however, which demonstrated the existence of a further isotropic liquid phase in addition to the aqueous and hydrocarbon micellar solutions usually obtained. The third phase was called "surfactant phase", and was observed at temperatures close to the phase inversion temperature (PIT).

This temperature was so designated because a transfer of the surfactant from the aqueous phase to the hydrocarbon phase occurred at this temperature. At low temperatures the surfactant caused a solubilization of hydrocarbon in the water which changed into a solubilization of water in hydrocarbon at temperatures higher than the PIT. Both solubilizations showed a maximum close to the PIT value, the high solubilization regions being restricted to a narrow temperature range.

Complementary studies of the solubilization of water in hydrocarbons and chlorinated hydrocarbons have been performed by *Kitahara* (12–14). His results further demonstrated the narrow temperature range for maximal solubilization and also the way this was changed by mixing different chain lengths.

The results by *Frank* and *Zografi* (15) on the solubilization of water in hydrocarbons after additions of various sulfo-succinates are also of interest in connection with microemulsions. The results showed the existence of a *second* solubilization phase when water was added to hydrocarbon solution of sulfosuccinates. The second solubilization phase was characterized as blue translucent.

In view of the importance of these phenomena we found an investigation of the combined interactions between water, surfactant and hydrocarbon to be of value in order to obtain a coherent picture of the phenomena. Our interest was focussed by *Shinoda* to the second solubilization region observed by *Frank* and further of using the surfactant phase to shed further light on the microemulsion phenomena.

Experimental

The nonionic surfactant was tetraethyleneglycol dodecyl ether (Nikkol) with a purity of more than

*) Permanent adress: The Technical University, Faculty of Chemical Technology, Szczecin, ul. Pulaskiego 10 (Polen).

98% according to gas chromatography test. The hydrocarbons were hexadecane, decane, hexane, benzene, and dodecylbenzene all used without further purification. The water was twice distilled.

Two of the components were weighed into an ampoule to which the third component was added stepwise. After each addition the sample was vibrated for a few seconds by means of a New Science *NV*, Vibromi *PS* and returned to the thermostat. The solubility area was marked as the beginning and end of the range where the solutions became transparent. For each region at least 20 determinations were made in this manner and a few samples were sealed and stored for 24 hours in order to establish the absence of long-term effects. Storage times in excess of 24 hours were avoided in view of the possibilities of bacterial degradation (16).

Results

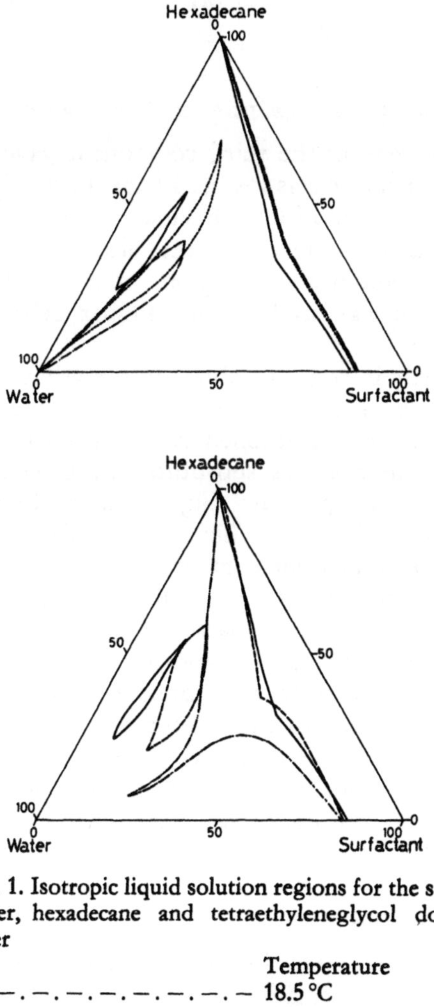

Fig. 1. Isotropic liquid solution regions for the system water, hexadecane and tetraethyleneglycol dodecyl ether

	Temperature
–.–.–.–.–.–.–.–.–	18.5 °C
- - - - - - - - - - - - - - -	20 °C
________________	25 °C
– – – – – – – – – –	30 °C
–..–..–..–..–..–..–	35 °C

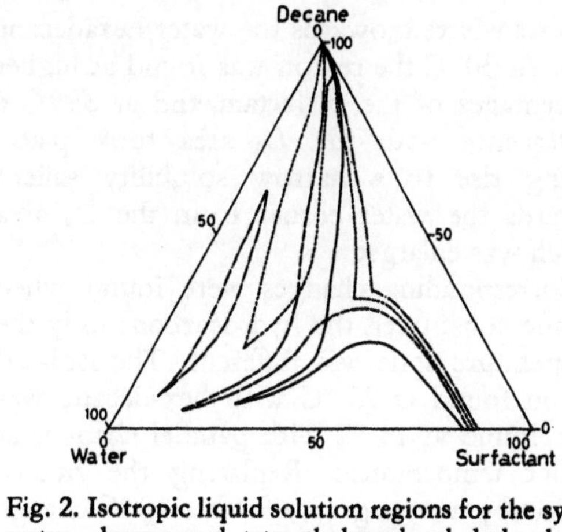

Fig. 2. Isotropic liquid solution regions for the system water, decane and tetraethyleneglycol dodecyl ether

	Temperature
–.–.–.–.–.–.–.–.–	15 °C
- - - - - - - - - - - - - - -	20 °C
– – – – – – – – – –	25 °C
________________	35 °C

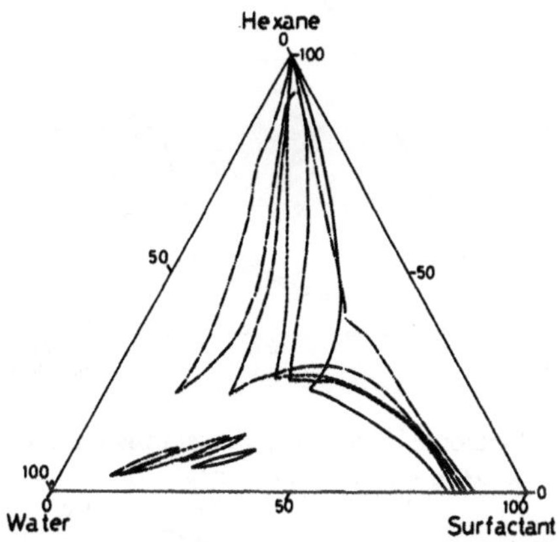

Fig. 3. Isotropic liquid solution regions for the system water, hexane and tetraethyleneglycol dodecyl ether

	Temperature
–..–..–..–..–..–..–	10 °C
–.–.–.–.–.–.–.–.–	15 °C
- - - - - - - - - - - - - - -	20 °C
– – – – – – – – – –	25 °C
________________	35 °C

Temperature effect on alkanes

Fig. 1 shows the solubility regions when the hydrocarbon was hexadecane. At 18.5 °C the isotropic liquid region extended from the water corner in a sector which ended at a water percentage of 40. A temperature increase to 20 °C caused a thinning of the sector which also reached further towards the hexadecane corner (max 68% $C_{16}H_{34}$). After a further increase in temperature to 25 °C the connection with the water corner was broken und the area of the region was reduced. The region was also transferred towards the water/hexadecane axis. At 30 °C the region was found at higher percentages of the surfactant and at 35 °C a coalescence with the L_2 area took place, giving rise to a narrow solubility salient towards the water corner from the L_2 area which was enlarged.

Corresponding changes were found when decane constituted the hydrocarbon; only the temperature scale was different. The isolated region found at 25 °C with hexadecane was now found at 15 °C with parallel changes at higher temperatures. Replacing the hydrocarbon by hexane gave a further 10 °C change of the temperature scale.

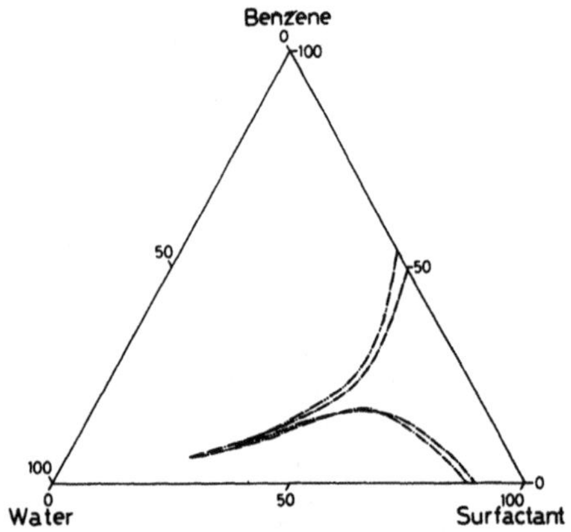

Fig. 4. Isotropic liquid solution regions for the system water, benzene and tetraethyleneglycol dodecyl ether

	Temperature
–..–..–..–..–..–..–	10 °C
–.–.–.–.–.–.–.–.–	15 °C

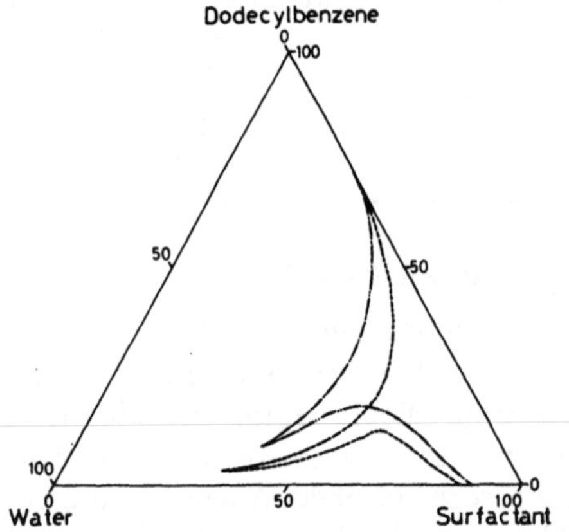

Fig. 5. Isotropic liquid solution regions for the system water, dodecylbenzene and tetraethyleneglycol dodecyl ether

	Temperature
–..–..–..–..–..–..–	10 °C
- - - - - - - - - - - - - - - -	20 °C

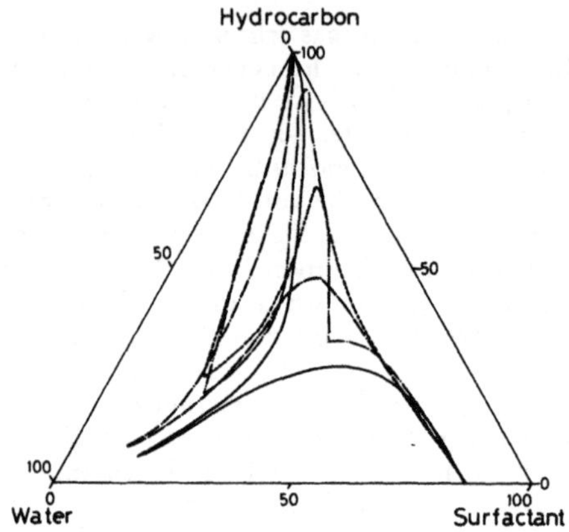

Fig. 6. Isotropic liquid solution regions at 20 °C for the system water, decane-benzene mixtures and tetraethyleneglycol dodecyl ether

	Benzene	Decane	
–..–..–..	0	100	
- - - - - - -	2	98	Weight percent
–.–.–.–.	7	93	
————————	10	90	

Influence of the aromaticity of the hydrocarbon

Benzene as the third component gave rise at temperatures as low as 10 and 15 °C (fig. 4) to a solubilization behaviour corresponding to that caused by the alkanes already at 30 °C. The solubilization region when dodecylbenzene was used showed similar behaviour (fig. 5).

Partly in order to obtain information about the influence of aromaticity, the solubility region was determined for hydrocarbon in which decane was substituted by benzene at constant temperature. Fig. 6 shows the continuous changes which took place with increasing benzene content.

Addition of two per cent benzene to the decane enlarged the narrow opening between the surfactant phase and the hydrocarbon solution and further additions gave rise to the characteristic solubility behaviour of decane at higher temperatures. At 10 per cent benzene the characteristic narrow salient pointing towards the water corner was formed.

Discussion

The present results demonstrated that the temperature sensitivity of solubilization by the nonionic surfactants (6–10) is not restricted to the transfer of the surfactant between the aqueous and hydrocarbon phases. All the diagrams displayed a pronounced temperature dependent shift of the solubility regions *parallel to the water-hydrocarbon axis*. This fact has an important bearing on the microemulsion phenomenon since the movements parallel to the water-hydrocarbon axis essentially imply changes of the emulsifier content of the solution.

The results show that the PIT value (6–10) not only denotes the temperature at which the surfactant is transferred from the aqueous phase to the hydrocarbon solution but that at this temperature the solutions require a minimum of emulsifier in order to solubilize large amounts of both water and hydrocarbons. This fact is well demonstrated by the curve in fig. 1, which shows the minimum concentration necessary to solubilize equal parts of water and hexadecane at various temperatures. The minimum at 25 °C is pronounced. The maximally permitted emulsifier concentration increases to 100 per cent when the temperature exceeds 30 °C since a coalescence with the L_2 region then takes place. The curve demonstrates the optimal emulsifier dosage required in order that the temperature range for solubilization may be the largest possible. The temperature sensitivity is clearly displayed by the maximal stability range of 6 °C at an emulsifier concentration of 13 per cent.

In general, designation of the surfactant phase as a microemulsion is justified; the emulsifier requirements are far below earlier reported concentrations for microemulsions. On the other hand, the present results do not support the opinion that the high mutual solubilization requires a special kind of emulsion. The fact that the solubilization areas where the minimal emulsifier concentrations are observed are not continuous extensions from the aqueous or hydrocarbon solutions but the direct connections to the aqueous solutions at temperatures below the PIT and that the coalescence takes place when the temperature is increased in excess of PIT, demonstrate that the surfactant phase is rather to be regarded as a narrow extension of the well known, common solubilized areas.

The connection with the second solubilization region observed by *Frank* and *Zografi* (15) is obvious; results similar to this are obtained by observing the solubility conditions at constant emulsifier concentration. Increasing the temperature causes the surfactant phase region to coalesce with the hydrocarbon solution but does not necessarily result in the second solubilization region completely merging with the hydrocarbon solution. This is illustrated in fig. 2. At 20 °C a second solubilization region will be observed when water is added to the hydrocarbon solution with between 7 and 50 per cent emulsifier. When the temperature is increased to 25 °C the two regions have coalesced and no secondary solubilization region is observed. It is, however, obvious from the present results that the first and second solubilization region form a continuous solubilization region already at 20 °C. The change when the temperature is increased to 25 °C consisted only in a widening of the channel between the two branches of the region.

The sensitivity of solubilization to the structure of the hydrocarbon was illustrated by the present results. A shortening of the chain length from sixteen to ten carbon atoms resulted in a general change of behaviour corresponding to a reduction in temperature by 10 °C. The solubilization region at 35 and 25 °C respectively for hexadecane corresponded to the regions at 25 and 15 °C for decane. It is essential to observe that the diagrams are drawn on a weight basis; the differences are solely due to the chain length and *not* to the total amount. A reduction of chain length to six carbon atoms also gave rise to a further reduction corresponding to a ten degree drop in temperature. In this case the narrow salient from the hydrocarbon/emulsifier solution towards the water corner which was characteristic for decane and hexadecane was split giving only a small isolated area of solution.

The presence of aromatic hydrocarbons, as expected, caused the same effect as a pronounced increase in temperature. Fig. 5 shows that a ten per cent addition of benzene to decane gave a change of solubilization matching a temperature increase of 5 °C. It is evident that this influence was not only dependent on the size of the benzene molecule: hexane showed a similar influence at higher concentrations throughout. Dodecylbenzene gave rise to solubilization similar to benzene in spite of the fact that the molecule has a size comparable with that of hexadecane. The difference is mainly to be found in the higher polarizability of benzene compared to that of the aliphatic hydrocarbons.

The connection between the surfactant phase and the hydrocarbon and aqueous micellar phases which was found in one investigation will in future investigations facilitate the determination of the structure of the surfactant phase. The results of such investigation will be reported.

Abstract

The regions for liquid isotropic solutions in the systems water, tetraethyleneglycol dodecyl ether and various hydrocarbons were determined at different temperatures. The earlier reported surfactant and second solubilization phases could be connected with the common aqueous and hydrocarbon based micellar phases. An increase in temperature resulted not only in a transfer of the emulsifier from the aqueous phase to the hydrocarbon phase but also in a regular shift of maximum water: hydrocarbon solubilizing compositions towards higher emulsifier: hydrocarbon ratios.

Acknowledgment

The authors are grateful to Professor *K. Shinoda* who suggested the investigation. The investigations were supported by Swedish Board for Technical Development.

Summary

The liquid isotropic solution regions in the phase diagrams of water, hydrocarbons and tetraethylene glycol dodecyl ether were determined by direct titration.

The results showed that at temperatures close to the Phase Inversion Temperature, PIT, separate regions appeared which contained large amounts of solubilized oil and water. The solutions are of importance for phenomena such as microemulsions, surfactant phases and second solubilization regions.

Zusammenfassung

Die Gebiete der isotropen Lösungen des Phasendiagramms Wasser, Kohlenwasserstoff und Tetraäthylen-glykoldodecyläther wurden bestimmt.

Die Ergebnisse zeigen, daß bei Temperaturen in der Nähe der „Phasen-Inversions-Temperatur" (PIT) getrennte Gebiete gefunden wurden, die große Mengen von Wasser *und* Kohlenwasserstoff enthalten. Die Lösungen dieser Art sind von großer Bedeutung, um Phänomene wie Mikroemulsionen, „Surfactant Phases" und „Zweite Solubilisation" zu verstehen.

References

1) *Hoar, T. P.* and *J. H. Schulman*, Nature **152**, 102 (1943).
2) *Schulman, J. H.* and *J. B. Montagne*, Ann. N. Y. Acad. Sci. **92**, 3661 (1961).
3) *Schulman, J. H., W. Stoeckenius* and *L. M. Prince*, J. Phys. Chem. **63**, 1677 (1959).
4) *Stoeckenius, W., J. H. Schulman* and *L. Prince*, Kolloid-Z. **169**, 170 (1960).
5) *Cooke, C. E.* and *J. H. Schulman*, in: Surface Chemistry p. 231. (Copenhagen 1965).
6) *Shinoda, K.* and *T. Ogawa*, J. Colloid and Interface Sci. **24**, 56 (1967).
7) *Saito, H.* and *K. Shinoda*, ibid, **35**, 359 (1971).
8) *Shinoda, K.*, ibid, **24**, 4 (1967).
9) *Shinoda, K.* and *H. Saito*, ibid, **26**, 70 (1968).
10) *Shinoda, K.* and *H. Kunieda*, ibid, **42**, 381 (1973).
11) *Gillberg, G., H. Lehtinen* and *S. Friberg*, ibid, **33**, 40 (1970).
12) *Friberg, S.*, J. Amer. Oil Chem. Soc. **48**, 578 (1971).
13) *Kitahara, A., T. Ishikawa* and *S. Taniusori*, J. Colloid Interface Sci. **23**, 243 (1967).
14) *Ishikawa, T.* and *A. Kitahara*, Kogyo Kagugaku Zasshi **71**, 96 (1968).
15) *Kou-no, K.* and *A. Kitahara*, J. Colloid Interface Sci. **37**, 469 (1971).
16) *Frank, S. G.* and *G. J. Zografi*, ibid. **29**, 27 (1969).

Authors' adress:

Prof. Dr. *S. Friberg*
Swedish Institute for Surface Chemistry
S-11468 Stockholm (Sweden)
Box 5607

Progr. Colloid & Polymer Sci. 56, 21–29 (1975)

4

Department of Physical Chemistry, Åbo Akademi, Åbo (Turku) (Finland)

The influence of solubilized N-Decanol on the stability of sodium N-Pentanoate (sodium valerate) micelles

P. Stenius and *L. A. Filén*

With 6 figures and 3 tables

(Received January 17, 1974)

Introduction

The influence of solubilizates on the size and shape of micelles formed by surfactants in aqueous solution, as well as the location of the solubilizate molecules in the micelles has been investigated extensively (1, 2). In these investigations, spectrophotometric methods have been of predominating importance. Few investigations of the thermodynamics of solubilization processes have been reported. As a result, little can be predicted quantitatively about the factors governing solubilization mechanisms and solubilizing capacities. In this investigation an attempt is made to give a quantitative description of the influence of a solubilizate, *n*-decanol, on the formation of micelles in sodium *n*-pentanoate (sodium valerate) solutions. The reasons for selecting this system are detailed below.

Thermodynamically, solubilization may be treated either using the "phase separation" approach or the "law of mass action" approach for the thermodynamics of micelle formation. In the former case, solubilization may be treated as a distribution equilibrium; the solubilizate is distributed between the aqueous phase (in which it is very slightly soluble) and the micellar phase (in which it often is very soluble) (3–6). In the "law of mass action" approach solubilization is treated as a formation of aggregates between the micelles and the solubilizate. This, besides the equilibrium between micelles and surfactant monomers, introduces new equilibria between solubilizing micelles and solubilizate monomers as well as between micelles containing solubilizate and micelles that do not (7). This approach would seem to be more realistic when the micelles formed by the surfactant are very small, as is the case in the present investigation. Thus, solubilization is described by a series of equilibria which may be written

$$sA^+ + qB^- + rD = A_sB_qD_r^{(q-s)-} \qquad [1]$$

if it is assumed that q surfactant ions B^- form micelles with s counter ions A^+ and r solubilizate molecules D.

In general, a thermodynamic description of solubilization could be given by determining the different equilibria of type [1] in a micellar solution (i.e., the values of s, q and r) as well as their equilibrium constants. In the case of surfactants forming large polydisperse micelles with a low critical micelle concentration, c.m.c., this, of course, is not experimentally feasible. Several methods of calculating average thermodynamic quantities have been published (e.g., ref. 7–10); they all lead to the calculation of the *Gibbs'* energy of micelle formation from the c.m.c. and thus imply that the c.m.c. should be sufficiently well defined to allow reliable calculations.

The ionic medium method

In a series of potentiometric and osmometric investigations (11–13), we have tried to show that the description of micelle formation in terms of equilibria of the type (1) (with $r=0$) can be experimentally verified for short-chain normal alkanoates. The equilibria in these cases can be investigated over wide concentration ranges and are simple enough to allow detailed calculations.

Since these salts are weak acids, the general equilibria investigated in a solution without added solubilizate may be written (charge signs are omitted from now on):

$$pH + sNa + qB = H_pNa_sB_q \quad [2]$$

If a large amount of swamping electrolyte, in this case sodium chloride, is added to give a constant ionic strength as the total concentration of B, C_B, is varied, the influence of sodium ions may be considered constant and the equilibria to be investigated are reduced to

$$pH + qB = H_pB_q \quad [3]$$

Moreover, since the ionic strength is constant and very high, it can be assumed that the activity coefficients are constant, i.e. the equilibria can be described in terms of concentrations. (The validity of these assumptions has been discussed in ref. 14). If a solubilizate D is added to the solutions the general equilibrium thus will be

$$pH + qB + rD = H_pB_qD_r \quad [4]$$

with the stability constant

$$\beta'_{pqr} = \frac{c_{pqr}}{[H]^p\,[B]^q\,[D]^r}\,. \quad [5]$$

The equilibria can be investigated experimentally by potentiometric determinations of the concentration of free hydrogen ions, $[H]$, as a function of the total concentration of hydrogen ions added, C_H, and the total concentration of anions, C_B and solubilizate, C_D (details will be given below). In previous investigations by this method it has been shown that aggregates of 4–5 anions with very similar stabilities are formed by sodium butyrate, pentanoate and hexanoate at ionic strength 3 M and 25 °C. The two latter salts also form micelles, the pentanoate micelle consisting of 11 and the hexanoatê micelle of about 17 anions. The hexanoate micelle is much more stable than the pentanoate micelle.

For this reason, the sodium pentanoate was considered a suitable substance for an investigation of the effect of a solubilizate on micelle formation. The micelle size and stability can be quite accurately determined and it is also possible to study the influence of solubilizate on the smaller aggregates formed. The solubilizate should be very slightly soluble in water, but solubilized to a reasonable extent. For this reason the weakly polar *n*-decanol was chosen.

Experimental

1. Materials

Sodium pentanoate was synthesized by neutralization of hot *n*-pentanoic acid (Fluka *puriss.* grade) with 1 M sodium hydroxide (Merck "Titrisol"). The purity was checked by titration in glacial acetic acid with perchloric acid; the molecular weight was within 0,25% of the theoretical value. Fluka *puriss.* grade *n*-decanol was used without further purification. The sodium chloride was Merck *pro analysi* grade; it was dried in vacuum at 110 °C. The water was doubly distilled and passed through an ion exchange resin; the conductivity was about $2\cdot10^{-6}$ ohm^{-1} cm^{-1}.

2. Potentiometric titrations

The association of the pentanoate anions was studied by investigating the hydrolysis of the aggregates formed as a function of the total pentanoate concentration C_B and decanol concentration C_D in a series of potentiometric titrations. These were all performed in solutions made 3 M in Na^+ by addition of $NaCl$. C_B and C_D were kept constant in each titration, the analytical excess of hydrogen ions, C_H, being varied by coulometric addition of hydroxide ions. The concentration of free hydrogen ions was measured with a hydrogen electrode in combination with the reference electrode.

$Ag, AgCl/3M\ NaCl$ (satd. with $AgCl/(3 - C_B)$ $M\ NaCl$, $C_B\ NaB)/$ where NaB denotes sodium pentanoate. The bridge solution thus was a neutral solution of similar composition to the solution titrated. The complete cell may be written
$Pt,\ H_2/H^+\ (3 - C_B)\ M\ NaCl,\ C_B\ NaB/$ reference electrode and its emf is given by

$$E = E^o_H - k\log[H] + E_j = \quad [6a]$$

$$= E^o_{OH} + k\log[OH] + E_j = \quad [6b]$$

$$= E^o_{OH} + k\log K_w - k\log[H] + E_j \quad [6c]$$

where E^o_H and E^o_{OH} are constant potentials, $k = RTF^{-1}\,ln\,10 = 59.156$ mV at 25 °C and E_j is the liquid junction potential.

The titration vessel and the system used to record the emf have been described elsewhere (15, 16). The titrations were performed automatically. A digital voltmeter with a print-out unit was used to measure the emf of the cell for 30 min after each addition of OH. The voltmeter was then turned off and the coulometer started, electrolytic addition of OH being performed for 30 min., followed by a new 30 min. recording period, etc. The emf values recorded, except for a few points in the immediate vicinity of the equivalence point, were stable within ±0,1 mV for 15–20 min.; the zero-point drift of the system was less than 0.2 mV/24 h. The highest current used in the electrolysis was 2 mA. The titration vessel was thermostated at 25 °C $\pm$ 0.05 °C.

In each titration, E°_{OH} was determined according to the method developed by *Biedermann* and *Sillén* (17). Known excess amounts of OH were added and the experimental quantity $E - k \log [OH] = E^{\circ}_{OH} + E_j$ for each addition then was plotted against $[OH]$ and extrapolated to $[OH] = 0$. This procedure invariably yielded straight lines and the intercept was taken as E°_{OH}; the slope gives a value for E_j. E_j was negligibly small in all titrations. The equivalence points were determined using Gran plots (18) from the E-values measured on the alkaline side of the equivalence point.

The $Ag, AgCl$ electrode was prepared by a method slightly modified from that of *Brown* (19). The hydrogen electrode was prepared according to *Bates* (20). The hydrogen gas was purified from oxygen and passed through two wash bottles containing 3 M $NaCl$ before entering the titration vessel. New $Ag, AgCl$ and hydrogen electrodes were prepared for each titration.

The discharge electrode was made of platinum. It had an area of ca. 4 cm² and was covered with platinum black by electrolysis in aqueous K_4PtCl_6. The anode was a silver plate heavily coated with $AgCl$ and immersed in 3 M $NaCl$.

3. Solubilization of decanol

The solubilization capacity for decanol of the three highest concentrations of sodium pentanoate investigated previously (12) (1.5, 2.1 and 2.8 M pentanoate) was checked by shaking weighed amounts of decanol with the solutions. The pH of the solutions was about 6.5. It was found that very small amounts of decanol were solubilized in 1.5 M pentanoate; in 2.1 M pentanoate the highest amount of decanol investigated was 0.1 mol decanol/mol pentanoate and in 2.8 M pentanoate the highest amount was 0.3 mol decanol/mol pentanoate.

Results and treatment of experimental data

Following *Sillén* and co-workers (21), the results of the titrations are described using the quantity

$$Z_{\text{exp}} = \frac{C_H - [H]}{C_B} \qquad [7]$$

Z_{exp} is the number of hydrogen ions bound per carboxylate ion. If the composition of all complexes formed in the solutions and their corresponding stability constants are known, theoretical values for Z may be calculated from (as is easily derived using eq. [5]):

$$Z_{\text{theor}} = \frac{\sum p [H]^p [B]^q [D]^r \beta'_{pqr}}{C_B} \qquad [8]$$

where the concentrations of free n-pentanoate ions, $[B]$ and decanol, $[D]$, may be calculated from

$$C_B = [B] + \sum q [H]^p [B]^q [D]^r \beta'_{pqr} \qquad [9]$$

$$CD = [D] + \sum r [H]^p [B]^q [D]^r \beta'_{pqr} \qquad [10]$$

the summations being taken over all p, q, r-values in the solutions.

By making guesses of the p, q, r-values and adjusting the corresponding β'_{pqr} values to fit Z_{theor} "as well as possible" to Z_{exp} one may find p, q, r-combinations which describe the experimental curves within the experimental uncertainty. This was done by the method of finding a minimum in a sum of square errors called *LETAGROPVRID*, developed by *Sillén et al.* (22, 23). In this procedure, the function minimized is

$$U = \sum_{i=1}^{N} (Z_{\text{theor}, i} - Z_{\text{exp}, i} + \delta Z)^2 \qquad [11]$$

where δZ is an adjustable systematic error in the Z values for each titration generally put $= 0$ in the first attempts to find the p, q, r-values; in the final tests it is thus possible to check the influence of systematic errors in the equivalence point (i.e., in C_H), in E°_{OH} (i.e., in $[H]$) and in C_B on the stability constants of the complexes. N is the number of experimental points. The standard deviation in Z is

$$\sigma(Z) = \frac{U}{N - n}$$

where n is the number of stability constants.

It was found that the calculations were quite insensitive to changes in r, which is probably due to the fact that the influence on the association of the decanol which does not bind hydrogen ions, is investigated by measuring the hydrogen ion activity – the influence will be "seen" in the experiments only indirectly as a change in the binding of hydrogen ions due to changes in the association of pentanoate ions as decanol is added. For this reason, it was assumed that the binding of hydrogen ions could be described by equilibria of the type given in eq. [3] and the effects of added decanol on p, q and β'_{pq} was then investigated by the *LETAGROPVRID* method.

The method of determining the standard potential E^o_{OH} described above implies that log $[OH]$ can be calculated directly from eq. [6b]. To calculate log $[H]$, (eq. [6c]), the ionic product of water K_w should be known. The value of K_w varies with the ionic strength and the composition of the solution. For this reason, the equilibria were written

$$pH_2O + q\,B = H_pB_q + p\,OH \qquad [12]$$

with the stability constants

$$\beta_{pq} = \frac{H_pB_q\,[OH]^p}{[B]^q}\,. \qquad [13]$$

Z_{theor} thus was calculated from

$$Z_{\text{theor}} = \frac{\sum p\,[B]^q\,[OH]^{-p}\,\beta_{pq}}{C_B} \qquad [14]$$

and $[B]$ from

$$C_B = [B] + \sum q\,[B]^q\,[OH]^{-p}\,\beta_{pq}\,. \qquad [15]$$

In all titrations reported here, $C_H \gg [H]$ and Z_{exp} was calculated from

$$Z_{\text{exp}} \approx \frac{C_H}{C_B}\,. \qquad [16]$$

Fig. 1 shows the values for Z_{exp} found for 2.1 M sodium pentanoate as increasing amounts of decanol are added. The corresponding curves for 2.8 M pentanoate are shown in fig. 2. It is seen that addition of decanol shifts the Z-curves towards lower values of ($-$ log $[H]$) in both cases. This is a direct indication of changes in the aggregation, similar to that found for additive-free sodium pentanoate

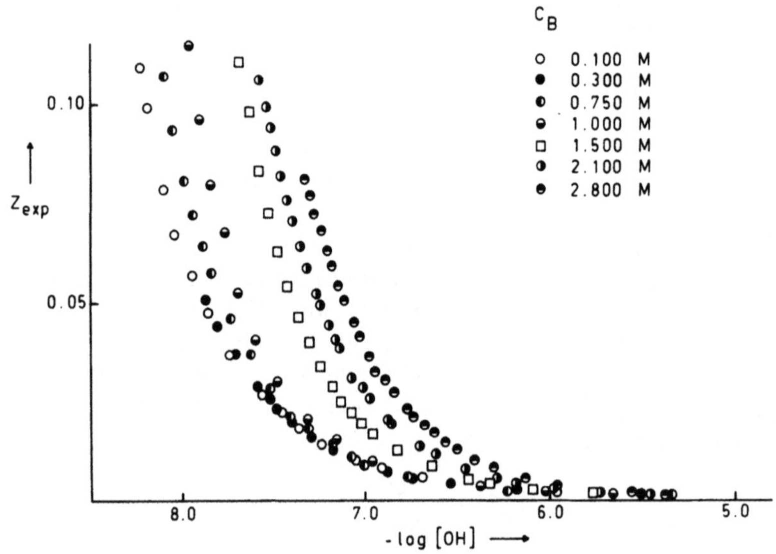

Fig. 1. The influence of added decanol on the association of 2.1 M sodium pentanoate at ionic strength 3 M. The number of hydrogen ions bound per pentanoate ion, Z_{exp}, is given as a function of $-$ log $[OH]$. The curves are shifted towards lower $-$ log $[OH]$ values as the amount of decanol increases, indicating an increase in the association.

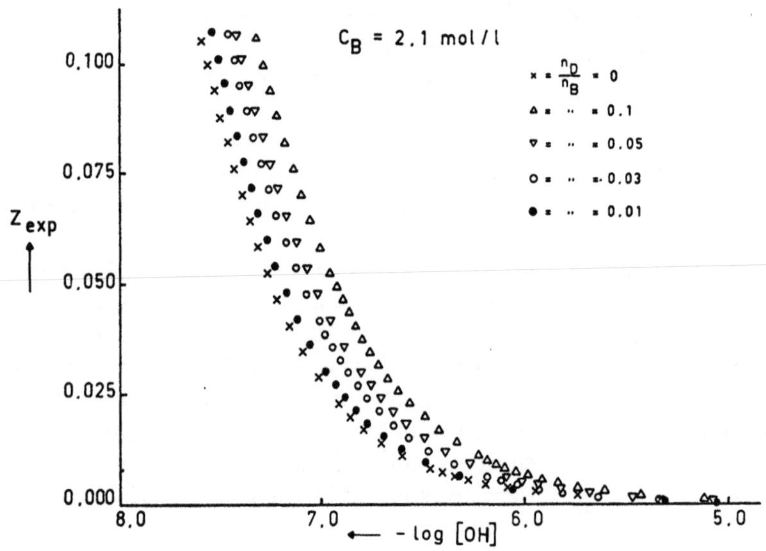

Fig. 2. As for figure 1; the total concentration of sodium pentanoate 2.8 M.

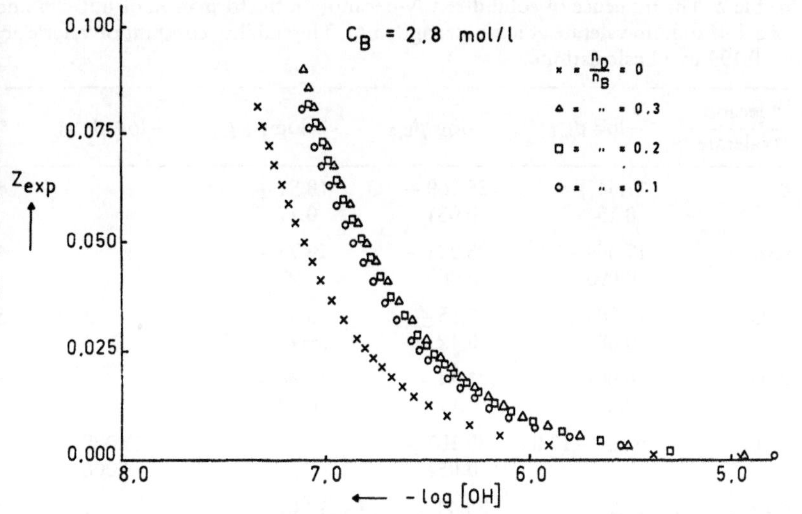

Fig. 3. The effect of increasing the concentration of sodium pentanoate on Z_{exp} as a function of $-$ log $[OH]$. For $C_B = 0.1$–1.8 M the curves are those reported in ref. 12; the curves for 2.1 M and 2.8 M pentanoate are results from renewed titrations. About 35% of the experimental points used in the calculations have been left out to avoid cluttering of points.

(fig. 3). As has been pointed out before (14), the method of calibrating the electrode system by determining E^0_{OH} for each titration, can lead to a shift of the Z_{exp} versus $-\log\,[OH]$ curves just because E^0_{OH} varies with C_B or C_D. However, in 2.1 M pentanoate, E^0_{OH} was 237.0 ± 0.5 mV for all C_D values. In 2.8 M pentanoate, there was a change in E^0_{OH} of about 10 mV from $C_D=0$ to the highest decanol concentrations; this is much too small to explain the shifts in Z_{exp}. In connection with the titrations in solutions containing decanol, titrations in 1.5, 2.1 and 2.8 M additive-free sodium pentanoate were renewed and similar calculations to those reported in ref. 12 were done using the new data. It was then found that the assumption that sodium pentanoate forms smaller aggregates with five anions gave a slightly better description of the experimental Z curves than the formation of aggregates with four anions reported previously. The standard deviation in Z decreases from 0.0050 to 0.0043. The precision attainable with the automatic titrator renders this a significant decrease; part of the data reported previously for high concentrations were obtained in manual titrations and would not have merited a distinction to be made. The systematic errors δZ for each titration were found to be 0.001–0.004. In view of the higher precision of the titrations reported here, the stability constants given in the last row of table 1 were accepted.

It was then investigated separately for 2.1 M and 2.8 M decanol whether addition of decanol changed the complex composition or the stability constants. In addition to the complex combinations given in tables 2 and 3, the following complexes were tried: HB_4, H_2B_4, H_3B_4, HB_6, H_2B_6, H_3B_6, B_{10}, HB_{10}, H_2B_{10}, HB_{12}, H_2B_{12}, H_3B_{12}, HB_{13}. The results show that

Table 1. The stability constants of the complexes formed in sodium valerate solutions at constant ionic strength 3 M ($NaCl$) and 25 °C. The first row gives complexes reported previously. In the following rows the stability constants for these complexes, as recalculated from renewed titrations at $C_B = 2.1$ M and 2.8 M, are compared to the complexes and stability constants that were found to give the "best" description of the renewed titrations

	$-\log\beta_{1,1}$	$-\log\beta_{2,4}$	$-\log\beta_{3,4}$	$-\log\beta_{2,5}$	$-\log\beta_{3,5}$	$-\log\beta_{1,11}$	$10^3\sigma(Z)$	$10^3\cdot U$
Stenius and *Zilliacus*	9.199 ± 0.022	17.077 ± 0.077	25.98 max.*) 25.69			12.22 max.*) 12.01	5.5	
This work	9.199 ± 0.022	17.062 ± 0.034	26.07 max.*) 25.83			12.80 ± 0.17	5.05	6.68
This work	9.199 ± 0.022			17.399 ± 0.058	25.262 ± 0.051	12.06 ± 0.20	4.32	5.03

*) max. = the standard deviation of the stability constant is larger than about 30% of the constant itself, and a maximum value of the constant is given instead of a standard deviation

Table 2. The influence of solubilized *N*-decanol on the formation of micelles and smaller aggregates in solutions of 2.1 *M* sodium valerate at ionic strength 3 *M*. The stability constant of valeric acid was given the value $-\log \beta_{1,1} = 9.199$ in all calculations

$n_{decanol}/n_{valerate}$	$-\log \beta_{2,5}$	$-\log \beta_{3,5}$	$-\log \beta_{2,11}$	$-\log \beta_{3,11}$	$10^3 \sigma(Z)$	$10^3 \cdot U$	Exptl points
0	17.64 ± 0.15	25.169 ± 0.051	18.81 ± 0.17		4.09	*)	219
0.01	17.369 ± 0.090	25.221 ± 0.060	18.78 ± 0.23		4.08		211
0.03	17.084 ± 0.083	25.35 ± 0.12	18.52 max. 18.29		5.05		220
0.05	16.997 ± 0.087	25.44 ± 0.17	18.36 ± 0.24		5.88		219
0.10	max. −17.0	25.107 ± 0.054		−25.028 ± 0.066	5.99	7.95	228
0.10	16.97 ± 0.16	25.44 ± 25.22	17.82 ± 0.17		7.41	12.30	228

*) A comparison of *U*-values for different decanol ratios is meaningless, since different experimental points were used in the calculations

a) the number of anions in the smaller aggregates and in the micelles does not change on addition of decanol

b) the stability of the smaller aggregates is slightly affected by the addition of decanol; $\beta_{2,5}$ increases somewhat and $\beta_{3,5}$ decreases.

c) the stability of the micellar aggregates increases on the addition of decanol

d) there is an increased tendency to bind hydrogen ions to the micelles as the amount of decanol is increased.

Table 3. The influence of stabilized *N*-decanol on the formation of micelles and smaller aggregates in solutions of 2.8 *M* sodium valerate at ionic strength 3 *M*. The stability constant of valeric acid was given the value $-\log \beta_{1,1} = 9.199$ in all calculations. The complexes that give the best description of experimental data are given in rows 1, 3, 5, and 8, respectively. Other values are given to show the magnitude of the effects in *U* and $\sigma(Z)$ as *p* and *q* are changed

	$n_{decanol}/n_{valerate}$	$-\log \beta_{2,5}$	$-\log \beta_{3,5}$	$-\log \beta_{1,11}$	$-\log \beta_{2,11}$	$-\log \beta_{3,11}$	$-\log \beta_{4,11}$	$10^3 \cdot \sigma(Z)$	$10^3 \cdot U$	Exptl points
1	0	17.445 ± 0.076	25.217 ± 0.047	12.31 ± 0.23				4.09	3.54	215
2	0	17.83 ± 0.17	25.155 ± 0.042		19.68 ± 0.16			4.32	3.94	215
3	0.10	17.23 ± 0.12	25.282 ± 0.088		18.90 ± 0.19			4.53		211
4	0.20	17.23 ± 0.15	25.277 ± 0.099		18.66 ± 0.16			4.83	4.86	211
5	0.20	17.88 ± max. 17.46	25.140 ± 0.075			25.83 ± 0.13		4.63	4.46	220
6	0.30	16.870 ± 0.066	25.81 max. 25.51	13.16 max. 12.63				6.92	10.39	220
7	0.30	17.01 ± 0.10	25.46 ± 0.19		18.74 max. 18.53			5.73	7.12	220
8	0.30	17.96 max. 17.47	25.133 ± 0.076			25.73 ± 0.12		4.54	4.48	220
9	0.30	17.020 ± 0.014	25.53 ± 0.23				33.98 max. 33.70	6.82	9.90	220

Discussion

It has been established long ago that there are substances which, when added to water, considerably increase the solubility of non-polar compounds in the aqueous solution without any obvious formation of micelles (25); the exact nature of this effect does not seem to be very well established. Among these so-called hydrotropic salts, the short-chain fatty acid salts have been investigated by *Durand* (26) and recently by *Markina, Kostova* and *Rehbinder* (27–29). It is found that the solubility of hydrocarbons increases with the concentration of sodium alkanoate for all salts with three or more hydrocarbon atoms in the alkyl chain. It is also found that hydrocarbons lower the c.m.c. of long-chain salts (27), while the increase in solubility for short-chain salts and for long-chain salts below the c.m.c. is ascribed either to the formation of smaller aggregates ("premicelles"), or to what is called a "hydrotropic effect".

The results presented above obviously shed some light on the possible differences in solubilization mechanism as the concentration of surfactant is increased from regions where no aggregates are formed to regions where the formation of micelles predominates. From the stability constants given in tables 1–3 the concentration of the different complexes can be calculated as a function of C_B and C_D. Fig. 4 shows the concentration of smaller aggregates and micelles as a function of C_B in additive-free sodium pentanoate solutions at $-\log [OH] = 7.5$. There is a region between 1 M and 2 M pentanoate where the formation of smaller aggregates predominates; at higher concentrations most of the substance bound into complexes is bound in micelles. A "c.m.c." can be very roughly defined at 1.2–2.0 M sodium pentanoate, in good agreement with the value 2.3 M given by *Hess, Philippoff* and *Kiessig* (30) for solutions without any added sodium chloride at 20 °C. From the solubilization experiments it can be concluded that almost no decanol is solubilized in regions where the smaller complexes predominate.

Figs. 5 and 6 show the change in concentrations of the different aggregates as decanol is added to 2.1 M and 2.8 M sodium pentanoate, respectively. In 2.1 M solutions there is a small increase in the concentration of smaller aggregates at very low concentrations of decanol; at higher concentrations, however, the concentration of the micelles increases. In 2.8 M solutions the concentration of smaller aggregates is considerably decreased as decanol is added. These curves directly show how the addition of decanol leads to an increased formation of micelles, i.e., a "lowering of the

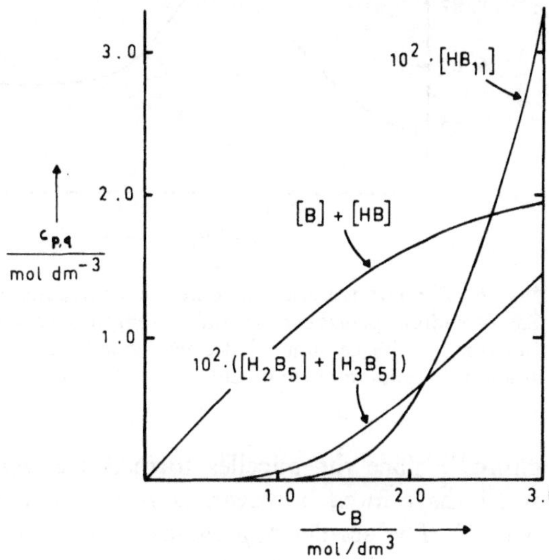

Fig. 4. The concentration of different aggregates in sodium pentanoate solutions as a function of the pentanoate concentration at constant ionic strength (3 M). The concentrations were calculated from the stability constants given in the last row in table 1. $-\log [OH] = 7.5$.

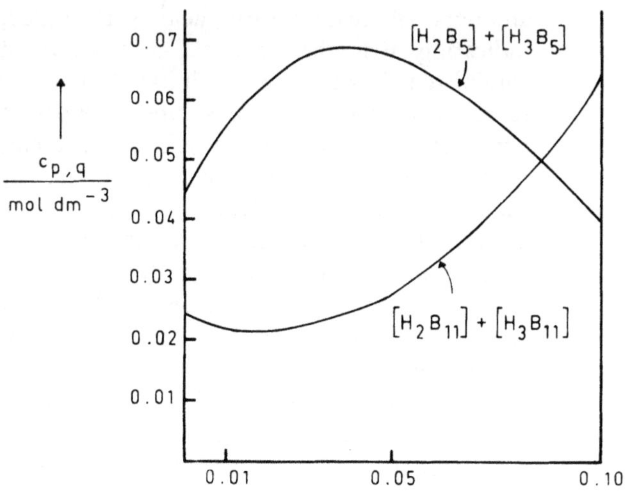

Fig. 5. The concentration of different aggregates in 2.1 M sodium pentanoate at ionic strength 3 M as a function of the amount of decanol added per unit amount of sodium pentanoate. $-\log [OH] = 7.5$.

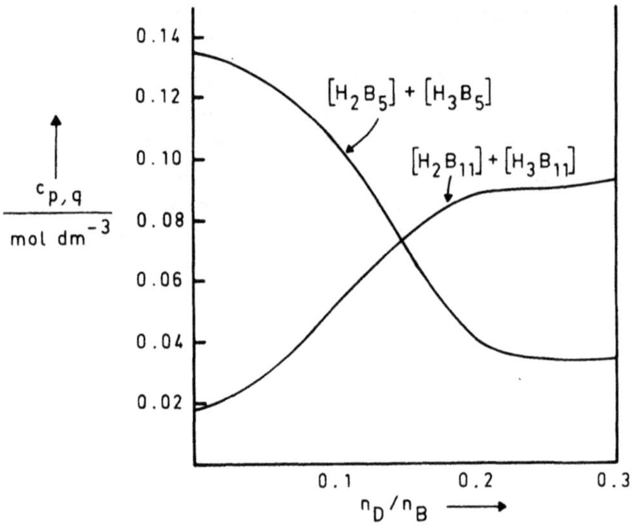

Fig. 6. The concentration of different aggregates in 2.8 M sodium pentanoate at ionic strength 3 M as a function of the amount of decanol added per unit amount of sodium pentanoate. – log $[OH] = 7.5$.

c.m.c."; since the micelles formed are very small the c.m.c., however, is not very well defined. The smaller aggregates do not seem to solubilize decanol to any significant extent. Since the smaller aggregates bind hydrogen ions much more strongly than do the micelles, the concentration of smaller aggregates increases if the solutions are made slightly acid. It was found that 1.5 M solutions saturated with decanol became turbid when very small amounts of hydrochloric acid was added, indicating that the very slight amount of solubilized decanol was precipitated. Thus, it seems that even in 1.5 M solutions, where a very small amount of micelles is present, decanol is solubilized only by them and not by the smaller aggregates. It may tentatively be asserted that the solubilization of non-polar compounds in these solutions does not take place by a formation of complexes with smaller aggregates, but through an increased formation of micellar aggregates. The size and concentration of these micelles is so small that it would be very difficult to investigate them by, for example, light scattering or viscosity measurements.

If it is assumed that all decanol added is bound to micelles, the largest pentanoate/decanol ratio in 2.1 M pentanoate corresponds to a micelle $H_3B_{11}D_3$. At $C_B = 2.8$ M the corresponding composition is $H_3B_{11}D_9$, a remarkably large proportion of decanol. If the sizes of the micelles are about the same in 2.1 M and 2.8 M solutions, the capacity of binding decanol should be independent of the total pentanoate concentration. This is in contradiction with the results presented above. However, the amounts of micelles and solubilized decanol in 2.1 M solutions are small, and the uncertainties in the stability constants (table 2) imply rather large uncertainties in the absolute concentrations of micelles, which, of course, also implies a large uncertainty in the value of r.

The investigation shows that it is possible to give a quantitative picture of the solubilization process in terms of stability constants for the aggregates formed. An even more complete description of the formation of these aggregates could be given by measuring the activity of the solubilizate. This can, as has been shown (31), be done by gas chromatography. Using such activities, a description in terms of the more realistic equilibria described by eq. [4] should be possible.

Acknowledgments

The authors thank professor *Ingvar Danielsson* for valuable discussions throughout the work.

Summary

The influence of a weakly polar solubilizate, *n*-decanol, on the stability constants and the aggregation numbers of smaller aggregates and micelles in aqueous sodium valerate (sodium *n*-pentanoate) solutions has been investigated by potentiometric titrations at 25 °C and constant ionic strength (3 M $Na(Cl)$). Under these conditions, the aggregation number of the micelles remains constant as decanol is added. However, the stability of the micelles is considerably increased. The smaller aggregates formed by valerate ions do not bind decanol to a measurable extent. At the highest concentrations of solubilized decanol investigated the micelles bind 0.5–0.8 mol decanol/mol valerate, implying a considerable increase in micellar weight even though the aggregation number of the pentanoate in the micelles remains constant on addition of decanol.

Zusammenfassung

Der Einfluß eines schwach polaren Solubilisats, *n*-Decanol, auf die Stabilitätskonstanten und die Aggregationszahlen kleinerer Aggregate und Micellen in wässerigen Natrium *n*-Valeratlösungen wurde mittels potentiometrischer Titrierungen bei 25 °C und konstanter Ionenstärke (3 M $Na(Cl)$) untersucht. Es wurde festgestellt, daß unter den genannten Bedingungen die Aggregationszahl der Micellen konstant bleibt, die Stabilität der Micellen jedoch durch die Zugabe von

Decanol gesteigert wird. Die durch die Valerationen gebildeten kleineren Aggregate binden kein Decanol in meßbarem Umfang. Bei den höchsten Konzentrationen des untersuchten solubilisierten Decanols binden die Micellen 0.5–0.8 Mol Decanol pro Mol Valerat, was eine beträchtliche Steigerung im Micellargewicht bewirkt, obwohl die Aggregationszahl des Valerats beim Zusatz von Decanol konstant bleibt.

References

1) *Elworthy, P. H., A. T. Florence* and *C. B. Macfarlane*, Solubilization by Surface-Active Agents and its Application in Chemistry and the Biological Sciences, (London 1968).

2) *Mysels, K.* and *P. Mukerjee*, Critical Micelle Concentrations of Aqueous Surfactant Systems, NSRDS-NBS **36** (Washington 1971).

3) *Nakayama, H., K. Shinoda* and *E. Hutchinson*, J. Phys. Chem. **70**, 3502 (1966).

4) *Hall, D. G.* and *B. A. Pethica*, in: *M. J. Schick*, (ed.), Nonionic Surfactants, p. 516 (New York 1967).

5) *Mukerjee, P.*, J. Pharm. Sci. **60**, 1531 (1971).

6) *Hall, D. G.*, Trans. Farad. Soc. **66**, 1351, 1359 (1970).

7) *Molyneux, P.* and *C. T. Rhodes*, Kolloid-Z. u. Z. Polymere **250**, 886 (1972).

8) *Phillips, J. N.*, Trans. Farad. Soc. **51**, 651 (1955).

9) *Overbeek, J. T. G.* and *D. Stigter*, Rec. Trav. Chim. **75**, 1263 (1956).

10) *Emerson, M. F.* and *A. Holtzer*, J. Phys. Chem. **69**, 3718 (1965).

11) *Danielsson, I.* and *P. Stenius*, J. Colloid Interface Sci. **37**, 264 (1971).

12) *Stenius, P.* and *C.-H. Zilliacus*, Acta Chem. Scand. **25**, 2232 (1971).

13) *Stenius, P.*, Acta Chem. Scand. (in press).

14) *Danielsson, I.* and *P. Stenius*, Trans. Royal Inst. Techn., Stockholm, Pure Appl. Chem. **34**, 81 (1972).

15) *Forsling, W., S. Hietanen* and *L. G. Sillén*, Acta Chem. Scand. **6**, 1034 (1957).

16) *Danielsson, I.*, Kemian Teollisuus **23**, 1081 (1966).

17) *Biedermann, G.* and *L. G. Sillén*, Arkiv Kemi **5**, 425 (1952).

18) *Gran, G.*, Analyst **77**, 61 (1952).

19) *Brown, R.*, J. Amer. Chem. Soc. **56**, 646 (1936).

20) *Bates, R.*, Determination of pH, p. 241 (New York, 1965).

21) *Sillén, L. G.*, Acta Chem. Scand. **10**, 186 (1956).

22) *Sillèn, L. G.*, Acta Chem. Scand. **18**, 1085 (1964).

23) *Sillén, L. G.* and *B. Warnqvist*, Arkiv Kemi **31**, 315, 341 (1969).

24) *Danielsson, I., A. Mäntylä, E. Kreutzman* and *P. Ritola*, Acta Acad. Aboensis, Math. Phys. **24**, 4 (1964).

25) *Winsor, P. A.*, Solvent Properties of Amphiphilic Compounds, p. 95, 119, (London, 1954).

26) *Durand, R.*, Compt. Rend. **223**, 898 (1946).

27) *Markina, Z. N., N. Z. Kostova* and *P. A. Rebbinder*, Doklad. Akad. Nauk. SSSR, Eng. Trans., **191**, 220 (1970).

28) *Kostova, N. Z., Z. N. Markina, P. A. Rebbinder* and *A. E. Kuz'mina*, Colloid J. USSR, Eng. Trans. **33**, 66 (1971).

29) *Kostova, N. Z.* and *Z. N. Markina*, Colloid J. USSR, Eng. Trans. **33**, 457 (1971).

30) *Hess, K., W. Philippoff* and *H. Kiessig*, Kolloid-Z. **88**, 40 (1939).

Authors' address:

Department of Physical Chemistry,
Åbo Akademi,
Porthansgatan 3–5,
20500 Åbo (Turku) 50, (Finland)

Progr. Colloid & Polymer Sci. **56**, 30–36 (1975)

5

Department of Physical Chemistry, Åbo Akademi, Åbo 50 (Finland)

Investigation of isotropic micellar solutions in the system sodium octanoate — N-Alcohol — water by small-angle X-ray scattering

B. Svens and M. Turpeinen

With 5 figures and 1 table

(Received January 17, 1974)

This report continues previous investigations of the association of alkali salts of short-chain alkyl carboxylates. In connection with these the structure of aqueous sodium octanoate (NaC_8) solutions has been investigated by small-angle X-ray scattering (1). It was then found that the scattering intensity curve was considerably changed by solubilized decanol in a way which indicates an increased micellar radius. In order to obtain a better description of the mechanism of solubilization of N-alcohols, a comparative investigation has now been made of the solubilization of pentanol, octanol and decanol, respectively, in the aqueous solution region of the system NaC_8—H_2O (L1-phase).

In addition, the so-called inverted micelles (2–4) in the other isotropic region of the NaC_8—H_2O-decanol system (L2-phase, decanolic solution) have been investigated. The regions of existence of the L1 and L2 phases are shown in fig. 1.

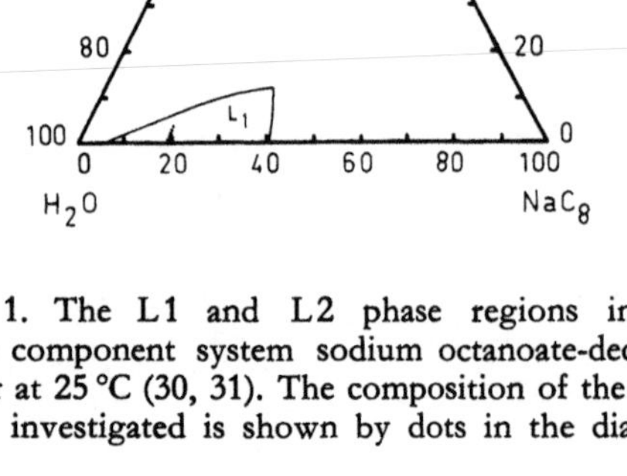

Fig. 1. The L1 and L2 phase regions in the three component system sodium octanoate-decanol-water at 25 °C (30, 31). The composition of the solutions investigated is shown by dots in the diagram

Experimental

Chemicals: Sodium octanoate was prepared by neutralization of octanoic acid (Fluka *puriss.*) with NaOH (Merck *p.a.*). The decanol and the pentanol were Merck *p.a.* quality; the octanol was redistilled once from Merck *purus* grade octanol. The water used was passed through an ion exchange resin and doubly distilled.

Apparatus: A Philips PW 1310/01/03 and PW 1320/00 X-ray power source fitted with a Cu fine focus tube was used. The scattering angle was measured with a Rigaku Denki small-angle goniometer. The detection system has been described previously (1).

Experiments: The sample solutions were prepared as described previously (1). The scattering curve was registered by scanning in steps of 0.03–0.07°, the time of measurement being 10 min for each step. In order to eliminate irrelevant parts of the scattered intensity, the scattering curve of a monomeric solution of sodium octanoate in water was subtracted from the curves recorded for L1-solutions. For this purpose, the scattering curve for 0.29 M NaC_8 (5) was used for systems containing decanol and the curve for 0.35 M NaC_8 was used for systems containing octanol or pentanol. 0.35 M NaC_8 corresponds to the critical micelle concentration, c.m.c., in pure aqueous solution (6–9). The scattering curve for pure decanol was subtracted from the curve recorded for L2-phase.

The resulting curves were evened out and subjected to slit corrections (10, 11).

All measurements were performed at (25 ± 1) °C. The uncertainties in the calculated parameters estimated from the uncertain factors in the experiments and the standard deviations calculated from the deviation between experimental and theoretical intensity curves (see below and ref. (15)) were: for the radii 0.5 Å, for the electron densities ϱ_{pol} and ϱ_2 0.001 and 0.005 e⁻/Å³, respectively.

Results and discussion

1. The phase region L1 in the systems NaC_8-decanol-H_2O, NaC_8-octanol-H_2O and NaC_8-pentanol-H_2O

All measurements were made at the lowest possible NaC_8 concentrations; it was therefore assumed that the influence of large aniso-dimensional aggregates and intermicellar interference effects could be disregarded.

In the calculations a relative intensity function $I(s)_{rel}$, which is valid for spherical particles, was used (1, 12). The function has been derived from the micelle model suggested by *Luzzati*. This model is characterized by regions with electron densities that can be chosen arbitrarily but are invariant within each region (13, 14):

$$\begin{aligned} I(s)_{rel} = \text{const.}\cdot[&R_{pol}^3 \varrho_{pol} \Phi(2\pi s R_{pol}) \\ &- R_{par}^3 \varrho_{pol} \Phi(2\pi s R_{par}) \\ &+ R_{par}^3 \varrho_{par} \Phi(2\pi s R_{par}) \\ &- R_{pol}^3 \varrho_{solv} \Phi(2\pi s R_{pol})]^2 \end{aligned} \quad [1]$$

where

R_{par} = the radius of the paraffin chain region (Å),

R_{pol} = R_{par} + thickness of the polar layer (Å),

ϱ_{par} = the electron density of the paraffin chain region ($e^-/Å^3$),

ϱ_{pol} = the electron density of the polar layer ($e^-/Å^3$),

ϱ_{solv} = the electron density of the solvent ($e^-/Å^3$).

$$\Phi(2\pi s R) = \frac{3(\sin 2\pi s R - 2\pi s R \cos 2\pi s R)}{(2\pi s R)^3}$$

$s = 2 \sin \theta/\lambda$

θ = half the scattering angle

λ = the wavelength of the scattered x-rays.

The electron density of pure hydrocarbons, 0.263 $e^-/Å^3$, was taken as the electron density ϱ_{par} of the paraffin chain region. The electron density of the solvent, ϱ_{solv}, was 0.336 $e^-/Å^3$ for 0.35 M NaC_8 and 0.335 $e^-/Å^3$ for 0.29 M NaC_8.

The theoretical scattering function $I(s)_{rel}$ was fitted to the experimentally determined scattering curve using a modified version of the computer program *Letagropvrid* (15). In this program the parameters R_{pol}, R_{par}, ϱ_{pol} and the proportionality constant in eq. [1] are adjusted to give the "best possible" description of the experimental curve, using a minimum in the least square deviation between theoretical and experimental points as a criterion. From this minimum deviations, standard deviations of the parameters are calculated. These give some indication of the real uncertainty in the parameters.

Fig. 2 shows some examples of the agreement between experimental and theoretical curves for different amounts of added octanol.

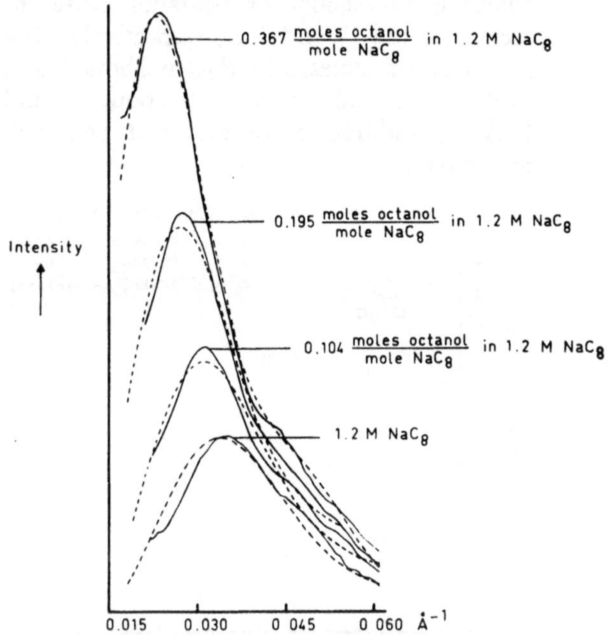

Fig. 2. The fitting of theoretical $I(s)$ curves to experimental $I(s)$ curves for the system NaC_8-octanol-water at different concentrations of octanol. ——— = experimental $I(s)$-curve; - - - - - = theoretical $I(s)$-curve

Older "X-ray diffraction" measurements were interpreted as showing that the thickness of disc-shaped micelles remains almost constant as N-alcohol is added (16). Similar opinions about a constant micellar volume still exist to some extent (17). Other investigations, however, show that the volume of the "mixed" micelles increases (e.g. 1, 18, 19).

As can be seen from fig. 3, there is a strong increase in the polar radius as decanol is added to 1.2 M NaC_8-water-decanol systems, while the increase in the paraffin chain radius is slower. The steady increase in the radii is caused by the swelling of the micelles as decanol is added to the palisade layer. At the same time, there is a decrease in ϱ_{pol} and a broadening of the polar layer. In the systems 1.2 M NaC_8-octanol-water and 1.2 M NaC_8-pentanol-water the changes in the micellar parameters were found to be smaller than in the corresponding systems containing decanol. An addition of about 0.2 mol decanol/mol NaC_8 causes an increase in the polar radius R_{pol} of about 10 Å, while similar molar additions of octanol or pentanol cause increases of only 7 and 1 Å, respectively. The simultaneous increase in R_{par} is about 2.5 Å as decanol is added while it is about 1.5 and 1 Å on addition of octanol and pentanol, respectively.

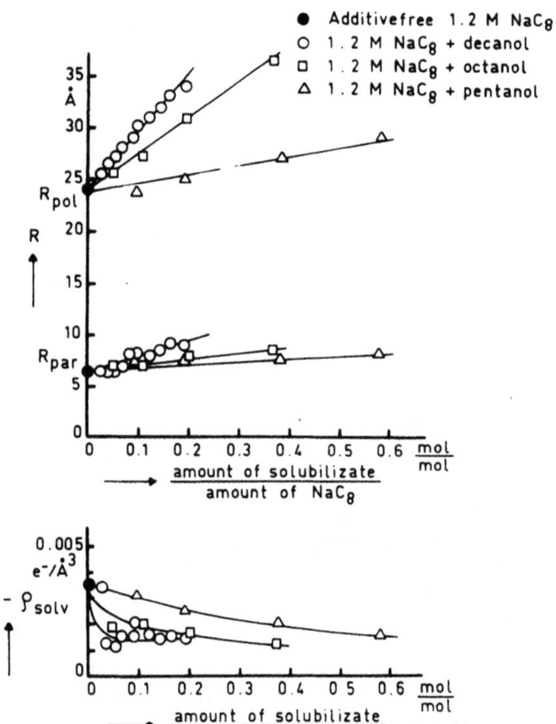

Fig. 3. The radii R_{pol} and R_{par} and the electron density differences $\varrho_{pol} - \varrho_{solv}$ as functions of the amount of added N-alcohol for the systems 1.2 M NaC_8-pentanol-water, 1.2 M NaC_8-octanol-water and 1.2 M NaC_8-decanol-water

The difference in electron density between the polar layer and the surrounding medium ($\varrho_{pol} - \varrho_{solv}$) decreases faster in the decanolic system than in the systems containing octanol or pentanol (even if the solubility of pentanol in water, 2.2 vol.%, at 25 °C, is taken into account, the value of $\varrho_{pol} - \varrho_{solv}$ is higher for the system containing pentanol than for the system containing decanol). The larger change in the electron density difference is connected with an increased broadening of the polar layer in the system containing decanol. The thickness of the polar layer in systems containing pentanol, on the other hand, remains constant when small amounts of pentanol are added. This indicates that the pentanol at low concentrations at least partly dissolves in the water.

The results are in accordance with previous investigations (2, 17, 20–22) which have shown that the hydrocarbon chain of the alkanol is directed towards the inner part of the micelle, while the hydroxyl group is located in the micellar surface. The present investigations, however, indicate that parts of the hydrocarbon chains proper of the decanol in octanoate solutions are not located in the central hydrocarbon part of the micelle; instead, they are layered in the palisade layer close to or on the surface. This seems reasonable, considering that the hydrocarbon chain of the decanol molecule is longer than that of the octanoate.

The markedly larger broadening of the polar layer as an alcohol with a long hydrocarbon chain is added gives added support to this view. $\varrho_{pol} - \varrho_{solv}$ decreases simultaneously. The octanol system shows a similar, but less marked effect, while on addition of pentanol, a very small increase is found in the polar layer at the highest concentrations.

The rapid decrease of $\varrho_{pol} - \varrho_{solv}$ in the solutions containing decanol possibly indicates an increased dissociation of counter ions from the micelles even at low concentrations of decanol. Conductivity measurements have also shown that polar solubilizates like N-alcohols liberate counter ions in micellar systems (*Heckman*; Na^+ (18), *Passinen* and *Ekwall;* Na^+ (19), *Hyde* and *Lawrence;* Br^- (23)).

2. The phase region L2 in the system NaC_8-decanol-H_2O

The isotropic phase L2, according to *Ekwall et al.* (5, 24–27) contains what is known as inverted micelles, with a core of water in contact with the hydrophilic groups of the amphiphile, the hydrocarbon chains being directed outwards towards the surrounding solvent. This description is now more or less accepted, but there are somewhat differing opinions as regards the structure of the micelles and the shape of the aqueous core.

Ekwall et al. (2, 28) emphasize the fundamental importance of the hydration of the alkali metal ions and the interaction with the alcohol in, for example, carboxylate-alcohol-water systems. The shape of the "inverted" micelles is considered to remain spherical up to relatively high concentrations of micellar substance (26).

According to the X-ray diffraction measurements of *Larsson* (29), the L2 range in systems of glyceride and water (12) is characterized by a lamellar structure. *Larsson* also proposes (29) that in the system CTAB-hexanol-water investigated by *Ekwall et al.* (25), the water forms cylindrical structures at low concentrations. As the concentration of water increases, there is a continuous transformation of the cylindrical structures to discs.

As a continuation of previous investigations it was decided to utilize the small-angle scattering of X-rays in the L2 phase formed by NaC_8, decanol and water.

In the interpretation of the scattering maxima obtained it was assumed that the micelles are spherical and inverted (*Ekwall* and others). The theoretical scattering function $I(s)$ used previously (eq. [1]) was modified in the following way to apply to the inverted micelles (see fig. 4):

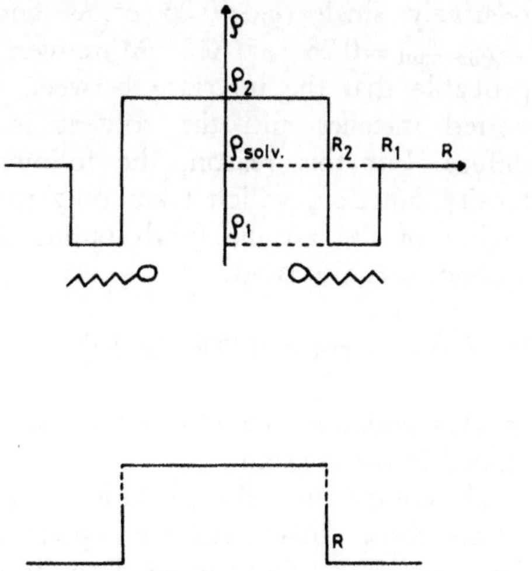

Fig. 4. Schematical descriptions of the radial electron density distribution in an inverted spherical micelle when (I) the difference in electron density between the solvent and the paraffin chain region is taken into account, (II) the difference is neglected

$$\text{I.}\quad I(s)_{rel} = \text{const.}\cdot[R_2^3\varrho_2\Phi(2\pi sR_2) - R_2^3\varrho_1\Phi(2\pi sR_2) + R_1^3\varrho_1\Phi(2\pi sR_1) - R_1^3\varrho_{solv}\Phi(2\pi sR_1)]^2. \qquad [2]$$

$I(s)_{rel}$ was fitted to the experimental $I(s)$-curve using the computer program *Letagropvrid* (15).

Table 1. Results from the L_2 phase

Concentration w %			Model I					Model II
NaC_8	decanol	H_2O	R_1/Å	R_2/Å	$\varrho_1/\frac{e^-}{Å^3}$	$\varrho_2/\frac{e^-}{Å^3}$	$\varrho_{solv}/\frac{e^-}{Å^3}$	R/Å
9.5	63.0	27.5	27.0	14.0	0.263	0.340	0.283	33
9.1	66.9	24.0						38
14.0	69.0	17.0	23.0	14.0	0.263	0.320	0.283	28
7.7	70.0	22.3						~40
12.5	70.0	17.5	21.0	14.0	0.263	0.305	0.283	26
15.0	70.0	15.0	21.0	14.5	0.263	0.305	0.283	26
16.0	70.0	14.0	21.0	15.0	0.263	0.300	0.283	25
10.7	74.3	15.0	20.5	14.0	0.263	0.300	0.283	24
5.0	80.5	14.5						~29
8.5	79.6	11.9	20.0	14.5	0.263	0.310	0.283	27
11.2	78.0	10.8	20.0	14.5	0.263	0.300	0.283	25
14.0	77.0	9.0	19.0	14.0	0.263	0.295	0.283	24

The difference in electron density between the paraffin chain region (if this is assumed to be a pure hydrocarbon) and the decanol is relatively small ($\varrho_1 = 0.26$ e⁻/Å³ and $\varrho_{solv} = \varrho_{decanol} = 0.28$ e⁻/Å³). Moreover, it is probable that the interface between the inverted micelles and the solvent is rather diffuse. For this reason, the following intensity function, which takes only the polar region of the micelle (with radius R) into account was also tried:

$$\text{II.}\quad I(s)_{rel} = \text{const.}\cdot[R^3\Phi(2\pi sR)]^2 \qquad [3]$$

In this equation the electron density is included in the constant.

The composition of the solutions investigated are shown in the three component phase diagram (30, 31) given in fig. 1 and in table 1. Distinct scattering maxima were obtained in the indicated area, where mainly spherical micelles have been found (32).

Curve-fitting according to model I: In general it was found to be somewhat more difficult to fit the theoretical scattering function to the experimental scattering curves of the L2 phase than to those of the L1 phase.

The results given in table 1 show that R_1 lies in the range (19–27) Å, while R_2 is almost constant (14–15) Å and ϱ_2 is (0.30–0.34) e⁻/Å³.

The total micellar radius R_1 increases as the concentration of water increases and as the concentration of decanol decreases. It should be observed that the carboxylate concentration seems to influence R_1 to a lesser extent. The size of the micelles thus seems to be primarily governed by the content of water. R_2 does not directly show the corresponding changes, but in spite of this a comparison of the values for the electron densities shows the increase in water content. Thus, ϱ_2 approaches the electron density of water ($\varrho_{H_2O} = 0.33$ e⁻/Å³) in the mixtures richer in water.

The thickness of the paraffin chain layer (R_1–R_2) is somewhat smaller than the length of the hydrocarbon chains of the carboxylate, in particular at low water contents. As a rule, the electron density of the aqueous region is somewhat lower than that of pure water. These circumstances indicate that the R_1 region partly or completely includes the first methylene group of the hydrocarbon chains. The carboxylate as well as the decanol molecules have pronouncedly lipophilic chains, and it thus seems very probable that the micelles have a rough surface structure with the hydrocarbon chains extending into the decanol.

"Curve-fitting" according to model II: The use of model II instead of model I implies that the electron density of the paraffine chain region of the micelle is assumed to be equal to that of the solvent. The scattering equation [2] then reduces to equation [3]. This approximation implies that parameters "broadening" the intensity maxima calculated from [2] are disregarded. For this reason, the curve-fitting has been limited to the range around the peak of the experimental scattered intensity ($s \simeq \pm 0.005$ Å⁻¹ from the peak, see fig. 5).

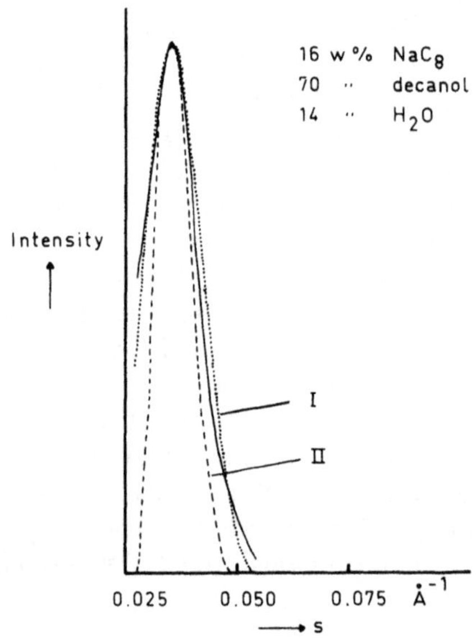

Fig. 5. The fitting of the theoretical $I(s)$-curve according to models I and II to the experimental $I(s)$-curve for the L2 phase in the system NaC_8-decanol-water

The radii found were throughout longer than those calculated according to model I. However, the changes in the radii reflect the variations in the water content quite clearly. As can be seen from table 1, the radius increases from 24 Å at 9 vol.% water to over 30 Å at a water content of about 25 vol.%. It is also seen that the radius decreases as the concentration of NaC_8 or the number of micelles increases.

In conclusion, it should be emphasized that the measurements in the L2 phase give values of the radii of spheres, whereas the real micelles may have very rough surfaces with the hydrocarbon chains protruding into the decanol. It is probably also difficult to define the exact borderline between the water and the polar region of the amphiphiles. When the radii found are discussed in connection with the electron densities it is, however, possible to draw conclusions regarding the micellar structure and the changes in it.

The large intermicellar regions have been disregarded in the discussion above, which has taken only the micelles proper into account. The total electron density of the micelles is implicitly expressed in terms of local electron densities and geometric micellar parameters. Since the electron density of a micelle is, in fact, connected to the electron density of the system as a whole, it may be concluded that the investigated local electron densities and micellar parameters are not quite independently variable; there is an implicit connection between them.

Acknowledgments

Professor *Ingvar Danielsson* and Dr. *Per Stenius* are thanked for valuable discussions. Parts of this work were sponsored by Neste Oy, Porvoo, Finland.

Summary

An X-ray investigation of aggregate dimensions and local electron densities in micellar systems of sodium octanoate (NaC_8) containing normal alcohols has been made, applying *Luzzati's* spherical micelle model. In aqueous solutions, the solubilizate being pentanol, octanol or decanol, the results indicate aggregates with thick polar layers, i.e., considerable hydration of the micelles. The thickness of the polar layer increases as the length of the hydrocarbon chain of the solubilizate increases.

In the case of systems with "inverted" micelles, that is, where water is solubilized in decanol, the influence of the difference in electron density between the solvent (decanol) and the hydrocarbon part of the aggregates has been investigated. The X-ray data indicate relatively large regions with a high electron density in the interior of the "inverted" micelles. This electron density lies between the electron density of water and that of decanol.

Zusammenfassung

Aggregatsdimensionen und lokale Elektronendichten micellarer Systeme von Natriumcaprylat wurden mittels Röntgenkleinwinkelstreuung – unter Verwendung des sphärischen Micellenmodells nach *Luzzati* – untersucht. Die erhaltenen Werte der wässerigen Lösung (L_1-Phase) mit Pentanol, Oktanol oder Dekanol als Solubilisat, deuten auf Aggregate mit breiten polaren äußeren Schichten, d.h. auf eine beträchtliche Hydratation der Micellen. Die Dicke der polaren Schicht nimmt mit wachsender Länge der Kohlenwasserstoffkette des Solubilisats zu.

Im ‚invertierten' micellaren Caprylatsystem – Wasser ist in Dekanol solubilisiert – wurde der Einfluß der Differenz der Elektronendichte zwischen Lösungsmittel (Dekanol) und dem Kohlenwasserstoffanteil des Aggregats untersucht. Die Daten aus der Röntgenkleinwinkelstreuung weisen auf ein relativ weitreichendes Gebiet mit hoher Elektronendichte im Innern der invertierten Micellen hin. Diese Elektronendichte in diesem Gebiet liegt zwischen dem jeweiligen Wert für Dekanol und Wasser.

References

1) *Svens, B.* and *B. Rosenholm*, J. Colloid Interface Sci. **44**, 495 (1973).

2) *Ekwall, P., I. Danielsson* and *P. Stenius*, in: *M. Kerker* (Ed.), Surface Chemistry and Colloids, MTP Intern. Rev. Sci., Phys. Chem. Ser. 1, vol. 7, p. 97 (London, 1972).

3) *Fowkes, F. M.*, in: *K. Shinoda* (Ed.), Solvent Properties of Surfactant Solutions, p. 65 (New York, 1967).

4) *Kitahara, A.*, in: *E. Jungermann* (Ed.), Cationic Surfactants, p. 289 (New York, 1970).

5) *Ekwall, P., L. Mandell* and *K. Fontell*, Molecul. Cryst. Liquid Cryst. **8**, 157 (1969).

6) *Mukerjee, P.* and *K. J. Mysels*, Critical Micelle Concentrations of Aqueous Surfactant Systems, p. 165 (Washington, 1971).

7) *Campbell, A. N.* and *G. R. Lakshminarayanan*, Can. J. Chem. **43**, 1729 (1965).

8) *Poland, D. C.* and *H. A. Sheraga*, J. Colloid Interface Sci. **21**, 273 (1966).

9) *Gustavsson, H.* and *B. Lindman*, J. C. S. Chem. Comm. 93 (1973).

10) *Schmidt, P. W.* and *R. Hight Jr.*, Acta Cryst. **13**, 480 (1960).

11) *Hendricks, R. W.* and *P. W. Schmidt*, Acta Phys. Austr. **26**, 97 (1967).

12) *Larsson, K.*, Z. Physik. Chem. (Frankfurt) **56**, 173 (1967).

13) *Reiss-Husson, F.* and *V. Luzzati*, J. Phys. Chem. **68**, 3504 (1964).

14) *Reiss-Husson, F.* and *V. Luzzati*, J. Colloid Interface Sci. **21**, 534 (1966).

15) *Sillén, L. G.* and *B. Warnqvist*, Ark. Kemi **31**, 315 (1968).

16) *Harkins, W. D., R. W. Mattoon* and *R. Mittelman*, J. Chem. Phys. **15**, 763 (1947).

17) *Emerson, M. F.* and *A. Holtzer*, J. Phys. Chem. **71**, 3320 (1967).

18) *Heckman, K.*, Kolloid-Z. **136**, 67 (1954).

19) *Passinen, K.* and *P. Ekwall*, Acta Chem. Scand. **9**, 1450 (1955).

20) *Winsor, P. A.*, Chem. Rev. **68**, 1 (1968).

21) *Lindblom, G., B. Lindman* and *L. Mandell*, J. Colloid Interface Sci. **44**, 400 (1973).

22) *Harkins, W. D* and *H. Oppenheimer*, J. Am. Chem. Soc. **71**, 808 (1949).

23) *Hyde, A. J.* and *A. S. C. Lawrence*, Proceedings 3rd International Congress on Surface Activity, **1A**, 21 (1960).

24) *Ekwall, P.*, J. Colloid Interface Sci. **29**, 16 (1969).

25) *Ekwall, P., L. Mandell* and *P. Solyom*, J. Colloid Interface Sci. **35**, 266 (1971).

26) *Ekwall, P., L. Mandell* and *K. Fontell*, J. Colloid Interface Sci. **33**, 215 (1970).

27) *Gillberg, G.* and *P. Ekwall*, Acta Chem. Scand. **21**, 1630 (1967).

28) *Lindman, B.* and *P. Ekwall*, Mol. Cryst. **5**, 79 (1968).

29) *Larsson, K.*, Chem. Phys. Lipids **9**, 181 (1972).

30) *Ekwall, P., I. Danielsson* and *L. Mandell*, Kolloid-Z. **169**, 113 (1960).

31) *Mandell, L.* and *P. Ekwall*, Acta Polytech. Scand. Chem. Met. Ser., **74 I**, 102 (1968).

32) *Ekwall, P.* and *L. Mandell*, Acta Chem. Scand. **21**, 1612 (1967; *Ekwall, P.* and *P. Solyom*, Acta Chem. Scand. **21**, 1619 (1967).

Authors' address:

B. Svens and *M. Turpeinen*
Dept. of Physical Chemistry
Åbo Akademi, Porthansgatan 3–5
20500 Åbo 50 (Finland)

Progr. Colloid & Polymer Sci. 56, 37–41 (1975)

6

The Swedish Institute for Surface Chemistry, Stockholm (Sweden)

The solubilization of water in polyoxyethylene alkyl ethers and the influence on the bonds between the ethers and fatty acids

By *Britta Lincoln*

With 9 figures

(Received January 17, 1974)

The polyoxyethylene alkyl ethers constitute an important part of the nonionic surfactants, which are the subject of intensive research and development efforts (1). The different structures involved and the thermodynamic changes which occur in aqueous solutions of nonionic surfactants have been investigated by *Corkill* et al. (2).

For the nature and the factors which determine the formation of microemulsions, the association conditions of surfactants have been shown to be an important factor (3, 4). Association conditions in systems of water–surfactant–organic compounds are also important to emulsion stability (5); it has been found that the presence of a liquid crystalline phase increases the stability of the emulsion to a marked degree (6).

The mechanism of this association and the locations of various solubilized molecules are however only slightly known. For this purpose molecular spectroscopy has been found to be a useful tool (7–9).

Certain ratios of nonionic surfactant and a carboxylic acid have been shown to give rise to a sharp increase in their water solubilizing capacity. This phenomenon has been related with the bonds between the two polar groups (10). The influence of added water on these bonds also in the solubilization of water in the pure polyoxyethylene alkyl ether have been studied by molecular spectroscopy.

Experimental

Materials

The octanoic acid was from Eastman, yellow labelled (~99 per cent/GC). Polyoxyethylene dodecyl ethers were from Nikkol, Japan. The purity tested out at more than 98% by gas chromatographic analysis. The ordinary water was doubly distilled and the heavy water from Norsk Hydro, 99.8%.

Treatment of the samples

The samples were prepared by mixing known amounts in ampoules which were sealed by fusion. If the following determinations did not take place within a few hours, the sealed ampoules were stored in the dark in order to minimize hydrolysis reactions.

Infrared spectroscopy

The infrared spectra were recorded at 20.0 ± 0.5 °C by means of a Perkin Elmer 337 spectrophotometer using a thermostated cell (RIIC VLT2). The cell was fitted with KRS-5 windows.

The spectra were thin film registrations with no solvent present. The path length was measured optically by means of the interference pattern before the cell was filled, and the spectra were afterwards transformed to a path length of 0.01 mm.

In the wave number region 1850–1600 cm^{-1} an external recorder was used to expand the wave number scale to ensure correct recording of the carbonyl stretching band.

The absorbance was calculated using the base line method in which the base line was fitted at specific wave numbers.

Nuclear magnetic resonance

The high resolution spectra were obtained on a Perkin Elmer R12B spectrometer equipped with the Variable Temperature Accessory. The temperature was controlled by using the relative chemical shift of the methanol-OH proton. The stability was ± 0.5 °C.

Chemical shifts are given relative to the methyl signal as an internal standard in order to avoid the use of bulk susceptibility corrections. Line widths were determined directly as half widths.

Results

The NMR spectrum of a pure polyoxyethylene alkyl ether consists of five signals: One from the methyl group protons, in this study used as an internal standard and not shown; one from the major part of the hydrocarbon chain methylene group protons, here called the $-CH_2-$ protons; one signal from the protons of the methylene group on the hydrocarbon chain placed closest to the first ether linkage, here called α-CH_2; one from the oxyethylene protons, here called the $-O-(CH_2)_2-O-$ protons and one from the terminal OH group, here called the –OH proton. When water was added, a rapid exchange of protons took place between the water protons and the proton of the terminal OH group of the ether, and one single resonance signal only was observed from these protons.

The infrared absorption bands of interest are those emanating from vibrations in the polar group. For the present investigation especially the anti-symmetric CO stretching vibration of the carboxylic group at 1715 cm^{-1} and the out-of-plane bending vibration of the OH group in carboxylic acid dimers at 940 cm^{-1} gave useful information.

Adding water to polyoxyethylene alkyl ethers

Fig. 1 shows the chemical shift of hexaoxyethylene dodecyl ether at 35 °C when water,

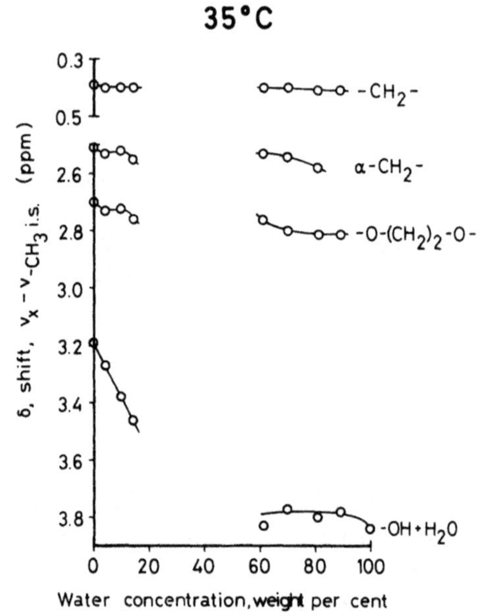

Fig. 2. The chemical shift of the signals of octaoxyethylene dodecyl ether at 35 °C when water was added

and at higher concentrations than 40% heavy water, was added. At this temperature there are liquid crystalline phases in the concentration range between 15 and 60% which are not possible to investigate with this technique.

The shift of the $-CH_2-$ signal was quite independent of the water content while the α-CH_2 and $-O-(CH_2)_2-O-$ signals showed a slow down-field shift. The $-OH + H_2O$ signal showed a strong down-field shift at the first

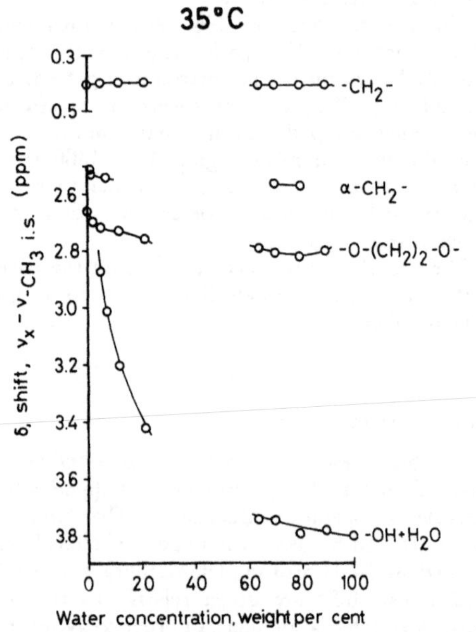

Fig. 1. The chemical shift of the signals of hexaoxyethylene dodecyl ether at 35 °C when water was added

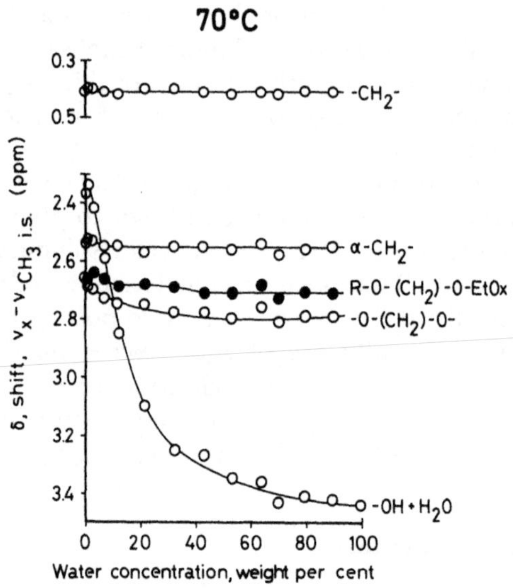

Fig. 3. The chemical shift of the signals of octaoxyethylene dodecyl ether at 70 °C when water was added

addition of water, but did not vary much with the water content at high concentrations.

Fig. 2 shows the same shifts for the octa-oxyethylene dodecyl ether as fig. 1 did for the ether with six oxyethylene groups. The shifts for the $-CH_2-$, $\alpha-CH_2$ and $-O-(CH_2)_2-O-$ signals were almost equal, but the $-OH + H_2O$ signal showed resonance at 0.4 ppm higher field than the lower homologue without water present. The difference diminished when the water concentration increased.

The shifts of the octaethylene dodecyl ether at 70 °C are shown in fig. 3. In addition to signals shown earlier, the signal of the protons of the oxyethylene group closest to the hydrocarbon chain may now also be observed. The signal of the $\alpha-CH_2$ protons showed a small up-field and then down-field shift at the first additions of water, and then became independent of the water concentration. The signal of the protons of the oxyethylene group nearest the hydrocarbon chain showed almost the same pattern, while the signal of the rest of the oxyethylene group protons shifted down-field until it was stabilized.

The $-OH + H_2O$ signal at 70 °C is shown in fig. 4 as a function of the proton fraction of water.

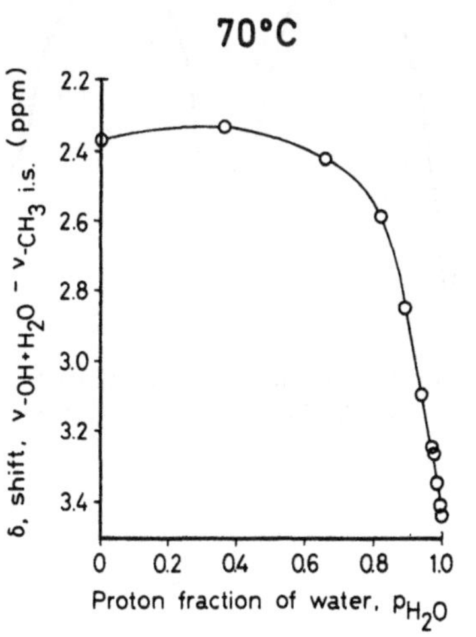

Fig. 4. The chemical shift of the coalesced signal from the terminal OH proton of octaoxyethylene dodecyl ether and the water protons at 70 °C as a function of p_{H_2O}, the proton fraction of water.

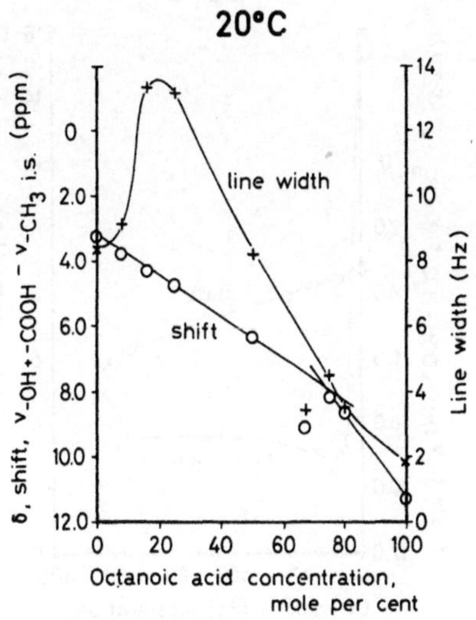

Fig. 5. The chemical shift (0) and the line width (+) of the coalesced signal from the carboxylic acid proton and the terminal OH proton of trioxyethylene dodecyl ether at 20 °C.

The influence of added water on the bonds between polyoxyethylene dodecyl ethers and octanoic acid

The shift and the line width of the coalesced signal of the acid proton of octanoic acid and the terminal OH group of the trioxyethylene dodecyl ether are shown in fig. 5. The signal function formed two straight lines meeting at approximately 75 mole per cent acid. The line width exhibited a maximum at about 20 mole per cent acid.

When 50 mole per cent water was added to the ether-acid samples the maximal line width appeared at the sample without acid, as shown in fig. 6.

The infrared absorption of the out-of-plane bending vibration of the carboxylic OH group at 940 cm^{-1} for the samples with trioxyethylene dodecyl ether and octanoic acid is given in fig. 7. With no water or only 10 mole per cent water present, the absorption increased rapidly when the acid concentration exceeded about 80 mole per cent. At higher water contents this increase started at lower acid concentrations but soon reached a maximum and decreased with further addition of acid.

Fig. 8 shows the shapes of the infrared absorption band of the anti-symmetric CO

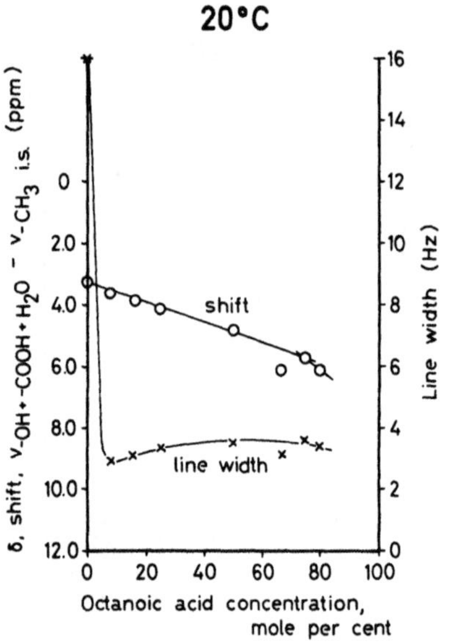

Fig. 6. The chemical shift (0) and the line width (×) of the coalesced signal from the water protons, the carboxylic acid proton and the terminal OH proton of trioxyethylene dodecyl ether with constant 50 mole per cent water present at 20 °C.

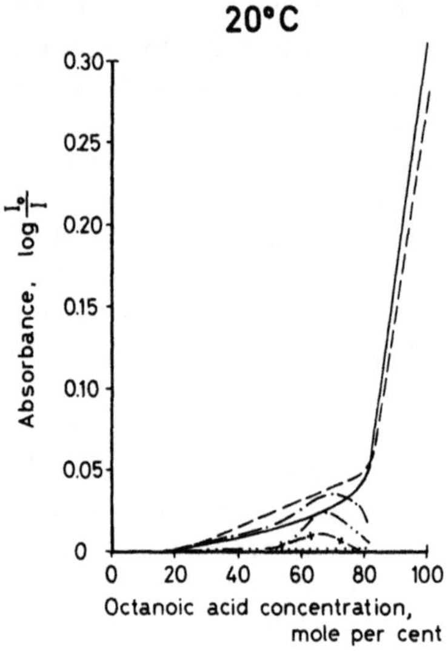

Fig. 7. The infrared absorbance at 20 °C of the out-of-plane deformation vibration of the acid dimers when trioxyethylene dodecyl ether was added to octanoic acid at different water concentrations.

———————— 0 mole per cent water
— — — — — 10 mole per cent water
–.–.–.–.–.–.–. 20 mole per cent water
—..—..—..—.. 30 mole per cent water
—+—+—+—+ 40 mole per cent water
.... 50 mole per cent water

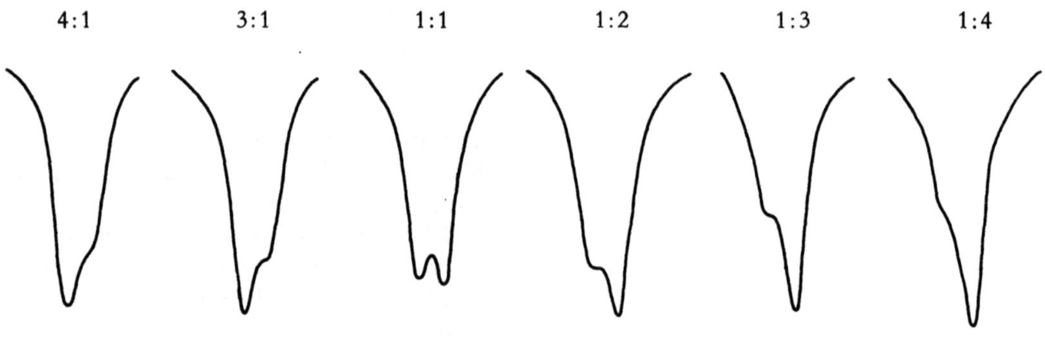

Fig. 8. The shape of the absorption band of the anti-symmetric stretching vibration of the carboxylic group varied with the ratio of trioxyethylene dodecyl ether to octanoic acid. The column notations show the molecular ratios of ether to acid. 20 °C

stretching vibration of the carboxylic group at 1715 cm^{-1} for different molar ratios of trioxyethylene dodecyl ether to octanoic acid. When the ether was added to the acid, a new absorption appeared at 1470 cm^{-1}, which dominated when the molecular ratio of ether to acid was sufficiently high.

When water was added to the samples, the adsorption at 1715 cm^{-1} increased at the expense of the absorption at 1740 cm^{-1}, as shown in fig. 9. This was the case both when the band at 1740 cm^{-1} was greater (upper series) and smaller (lower series) than the one at 1715 cm^{-1}.

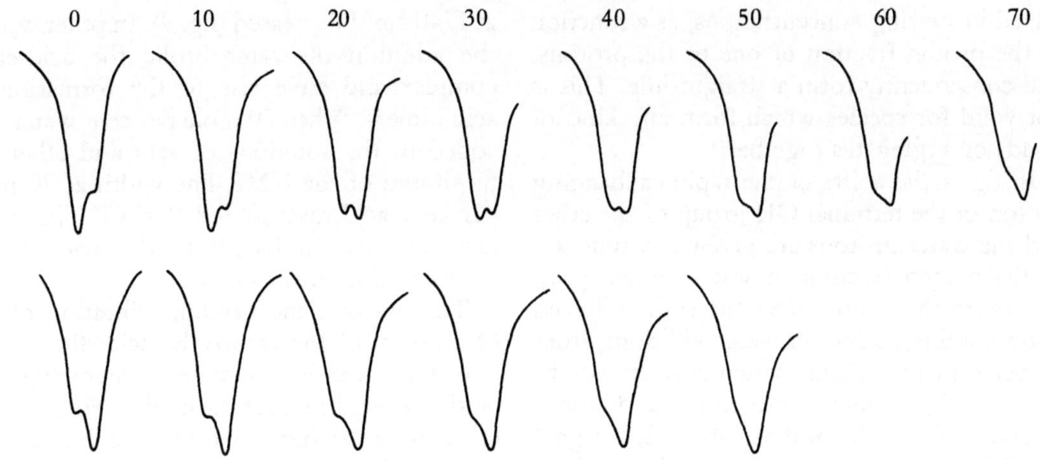

Fig. 9. The shape of the absorption band of the anti-symmetric stretching vibration of the carboxylic group when water was added to samples with trioxyethylene dodecyl ether and octanoic acid at different molecular ratios of ether to acid, upper series 3:1 and lower series 1:2. The column notations show the mole percentage of water. 20 °C

Discussion

Adding water to polyoxyethylene alkyl ethers

The addition of water to polyoxyethylene dodecyl ethers did not have any influence on the methylene chain more than on the methyl group. This is shown in figs. 1—3, where the shift of the $-CH_2-$ protons relative to the methyl group was quite independent of the water concentration, at both 35 and 70 °C and for different lengths of the oxyethylene chain. This indicated that the water was not located in the vicinity of the hydrocarbon chain at any concentration.

The protons of the oxyethylene chain in all three figures showed a down-field shift when water was added, which stopped at high water concentrations. A more hydrophilic environment and an increased hydrogen bonding cause shift changes to a lower field. The down-field shift of the $-O-(CH_2)_2-O-$ protons accordingly showed the increased water concentration around the oxyethylene chain.

At 35 °C the α-CH_2 protons experienced a contact with the water comparable with the oxyethylene chain. A small difference was however observed. The major down-field shift appeared to come at slightly higher water concentrations. The water, added to a polyoxyethylene alkyl ether, was consequently equally distributed among all the groups in the oxyethylene chain, but did not completely surround the ether linkage closest to the hydrocarbon chain until the water concentration was high.

At 70 °C the pattern was somewhat different. Even at small water concentrations the shift of the protons of the oxyethylene group closest to the hydrocarbon chain separated from the shift of the other oxyethylene groups. Both the α-CH_2 signal and the signal of the oxyethylene group closest to the hydrocarbon chain showed a small maximum at the first water additions. For further additions of water the α-CH_2 was independent of the water ocncentration, while the oxyethylene group showed a small dependence. This indicated that the added water was located at the end of the oxyethylene chain. If an aggregation of micelles takes place, the ether OH group will primarily be occupied by water and the oxyethylene group closest to the hydrocarbon chain will not be involved with the OH groups, hence a decreased hydrogen bonding. This mechanism could explain the initial up-field shift.

The α-CH_2 signal was independent of the water concentrations, which further supports the postulate that the water and the end groups of the oxyethylene chain form the hydrophilic environment of an aggregation.

When the exchange between different protons is so rapid that separate signals are not detectable, the observed chemical shift is a weighted average for the protons present (11). The shift of two different exchanging protons, mixed in varying concentrations, as a function of the proton fraction of one of the protons, will consequently form a straight line. This is not valid for species which form any kind of bonds or aggregates together.

In fig. 4 the shifts of the rapid-exchanging proton of the terminal OH group of the ether and the water protons are given as a function of the proton fraction of water. When p_{H_2O} was more than about 0.87 the shift followed a straight line, so further water additions, from 10 per cent by weight, which corresponds to $p_{H_2O} = 0.87$, simply represented a dilution.

The up-field shift of the $-OH + H_2O$ signal at 70 °C compared with the shift at 35 °C was the quite normal temperature dependence of the water signal, about 0.5 Hz/°C. The difference between the hexa- and octaethylene dodecyl ethers, when both were studied at 35 °C, could not however be explained in this way. For the range 0 to about 15% water, where all the polyoxyethylene alkyl ethers have isotropic liquid phases, the shift of the hexaethylene ether was representative for the ethers with two to six oxyethylene groups. The difference appeared then when the oxyethylene chain increased in length from six to eight groups. The only likely explanation of this higher resonance field of the terminal OH proton of the octaoxyethylene ether than of the lower homologues was a different conformation, which gave a more hydrophobic environment of the proton.

The influence of added water on the bonds between polyoxyethylene dodecyl ethers and octanoic acid

According to an earlier investigation (10) the tetraoxyethylene dodecyl ether and octanoic acid formed a complex containing 2 acid and 1 ether molecules, and at about 10 mole per cent acid there was an influence on the conformation of the ether which gave a broadening of the NMR signal. Trioxyethylene dodecyl ether also showed these effects but at slightly higher acid concentrations (fig. 5). The split of the infrared antisymmetric CO stretching vibration in fig. 8 also showed that acid molecules both form dimers with each other, (the band at 1715 cm^{-1}) and are bound to the ether (the band at 1740 cm^{-1}) (10).

Added water caused an increase of the absorption at 1715 cm^{-1}, while the absorption at 1740 cm^{-1} decreased (fig. 9). In other words, the addition of water broke the acid-ether complex and gave rise to the formation of acid dimers. When 50 mole per cent water was added to the solutions of acid and ether the maximum of the NMR line width at 20 mole per cent acid disappeared (fig. 6). The water eliminated the influence of the acid on the conformation of the ether.

The out-of-plane bending vibration of the OH group of the carboxylic acid dimers was also influenced by the water content (fig. 7). With no water present the conformation of the ether with 20 mole per cent acid disturbed the out-of-plane bending vibration of the acid and the infrared absorption band was absent. Further additions of acid to the ether caused the absorption to increase slowly, occasioned by the dilution between the conformation with 20 mole % acid and the complex with about 80 mole % acid, in which some of the acid molecules were bound to each other as dimers (fig. 8). When all the ether was bound to the acid in the complex, further additions of acid caused the absorption to increase markedly.

Ten mole per cent water did not change this pattern. With more water present, so much acid was relieved that the marked increase of the absorbance started at a lower acid content.

In spite of increasing acid content the out-of-plane bending absorbance did not continue but decreased. Since the anti-symmetric stretching vibration did not decrease, the effect on the out-of-plane bending vibration must be a steric obstacle by the ether and the water to the carboxylic group.

Twenty mole per cent water, at which this effect is obvious, corresponds only to about 2 per cent by weight. In other words, the marked influence of water on the bonds between polyoxyethylene dodecyl ethers and octanoic acid occurs at low water concentrations.

Acknowledgement

Dr. *Stig Friberg* is heartily thanked for initiating this investigation and for many fruitful discussions.

Summary

The influence of added water on polyoxyethylene dodecyl ethers was investigated by NMR and IR.

At 35 °C the water was equally distributed among all the groups in the oxyethylene chain, but did not completely surround the ether linkage closest to the alkyl chain until the water concentration was high. At 70 °C the ether formed aggregates as soon as water was added, so that the water never penetrated to the innermost ether linkage.

Water added to solutions of trioxyethylene dodecyl ether and octanoic acid containing a complex of about 3 acid and 1 ether molecules caused a removal of the acid from the complex.

Zusammenfassung

Der Einfluß von Wasserzugabe auf Polyoxyäthylendodecyläther wurde mittels NMR und IR untersucht. Bei 35 °C war das Wasser gleichmäßig zwischen allen Gruppen der Oxyäthylenkette verteilt, die Ätherbindung nächst der Alkylkette wurde jedoch erst bei hoher Wasserkonzentration von diesem vollständig umgeben. Bei 70 °C bildete der Äther bei Wasserzugabe sofort Aggregate, so daß das Wasser nie bis zur innersten Ätherbindung durchdrang.

Wasser, das zu Lösungen von Trioxyäthylendodecyläther und Caprylsäure – die einen Komplex aus etwa drei Säure- und einem Äthermolekül bildeten – zugegeben wurde, verursachte einen Ausschluß der Säure aus dem Komplex.

References

1) *Schick, M. J.* Nonionic Surfactants (New York, 1967).

2) *Corkill, J. M.* and *J. F. Goodman*, Advan. Colloid Interface Sci. **2**, 297 (1969).

3) *Gillberg, G., H. Lehtinen* and *S. Friberg*, J. Colloid Interface Sci. **33**, 40 (1970).

4) *Shinoda, K.* and *H. Kunieda*, J. Colloid Interface Sci. **42**, 381 (1973).

5) *Shinoda, K.* and *H. Saito*, J. Colloid Interface Sci. **30**, 258 (1969).

6) *Friberg, S., L. Mandell* and *M. Larsson*, J. Colloid Interface Sci. **29**, 155 (1969).

7) *La Force, G.* and *B. Sarthz*, J. Colloid Interface Sci. **37**, 254 (1971)

8) *Sarthz–Lincoln, B.* and *S. Friberg*, Proc. VIth Int. Congr. Surf. Act. Subst., Zürich, 1972.

9) *Lincoln, B.*, To be published.

10) *Lincoln, B., S. Friberg*, and *S. Gravsholt*, Kolloid-Z. u. Z. Polymere (In press).

11) *Bovey, F. A.*, Nuclear Magnetic Resonance Spectroscopy, p. 187 (New York, 1969).

Author's address:

Frau Dr. *Britta Lincoln*
AB Volvo-Bergstagsverken Fack
S-73101 Köping 1
Schweden

Für die Schriftleitung verantwortlich: Für Originalarbeiten Prof. Dr. F. H. Müller, 3550 Marbach b. Marburg/L. und Prof. Dr. Armin Weiss, 8000 München 2
Dr. Dietrich Steinkopff Verlag, 6100 Darmstadt, Saalbaustraße 12
Herstellung: Meister Druck, 3500 Kassel

Mitarbeiter-Bedingungen · Note to Contributors

Originalbeiträge sind an die folgenden Herren zu senden:

Arbeiten aus dem Bereich der Polymerforschung an

Arbeiten aus dem Bereich der Kolloidchemie und Biochemie an

The authors are requested to submit their **manuscripts** to the following Editors:

Contributions on Polymer Science to

Prof. Dr. F. H. Müller (Direktor des Univ.-Instituts für Polymere, Marburg/Lahn) 3550 Marbach bei Marburg/Lahn, Schulstr. 26;

Contributions on Colloid Science and Biochemistry to

Prof. Dr. A. Weiss (Institut für Anorganische Chemie der Universität München) 8000 München 2, Meiserstr. 1;

Die Zeitschrift veröffentlicht nur angeforderte Originalbeiträge zu jeweils einem bestimmten Thema pro Band.

This journal will publish original contributions only on request by the editors covering the special scope of each volume.

Manuskripte sollen in zweifacher Ausfertigung eingereicht werden. Ihr Eingang wird umgehend bestätigt. Ihr Inhalt muß unveröffentlicht sein. Die Verantwortung für den Inhalt liegt bei den Autoren. Publikationssprachen: Deutsch, Englisch oder Französisch. Jedem Manuskript ist eine Zusammenfassung in deutscher und englischer Sprache beizugeben. Die Typoskripte müssen einseitig und weitzeilig geschrieben sein. Abbildungen sind mit Legenden zu versehen und als klischierfähige Vorlagen einzureichen, wobei die Beschriftung auf einem transparenten Deckblatt anzubringen ist. Formeln bitte deutlich schreiben, insbesondere griechische Buchstaben und Indices! Die Zahl der Abbildungen und Tabellen ist auf das unbedingt Notwendige zu beschränken. Für Literaturangaben gelten die international üblichen Regeln. Die Literatur ist am Schluß der Arbeit zusammenzufassen. – Anstelle eines Honorars erhalten die Autoren insgesamt 75 Sonderdrucke kostenlos, weitere Exemplare auf ausdrücklichen Wunsch gegen Berechnung. – Kosten für nachträgliche Autorkorrekturen, soweit es sich um Textergänzungen in der Druckfahne handelt, werden dem Autor in Rechnung gestellt. – Ausführliche Sonderdrucke der geltenden Mitarbeiterbedingungen sind kostenlos beim Verlag erhältlich. – Nicht den Richtlinien entsprechende Manuskripte werden zurückgesandt.

Manuscripts should be submitted in duplicate and should contain original work as yet unpublished elsewhere. Their receipt will be acknowledged promptly. Authors are fully responsible for the contents of their contributions. Publications languages: English, French or German. Each manuscript should include a summary in English and German. All manuscripts should be double-spaced, typed on one side only. Illustrations and drawings should be made carefully, with India ink on white drawing paper, blue tracing linen or coordinate paper ruled in blue only. Lettering at the sides of graphs may be pencilled in and will be typeset. Legends must accompany the drawings, Formulas, symbols and Greek letters should be carefully made and annotated and subscripts and superscripts clearly shown. The number of figures and tables should be held to a minimum. The list of references should be written on a separate page. It is recommended that abbreviation of the titles of the Journals be made in conformity with Chemical Abstracts (see List of Periodicals, 1961).
Authors will receive 75 reprints of their contribution free of charge and may order an additional number at cost. Authors making elaborate alterations and additions in proof will be required to bear the costs thereof. More detailed instructions to the authors can be obtained from the publisher free of charge. Manuscripts which do not conform with the above guidelines will be returned to the authors.

Geschäftliche Bedingungen · Note to Subscribers

Erscheinungsweise:
Zwanglos nach Bedarf in Bänden verschiedenen Umfangs.

Frequency of Publication:
Irregularly in volumes of different size.

Bezugspreis dieses Bandes:
DM 28,– plus Porto.
Bezieher der Kolloid-Zeitschrift & Zeitschrift für Polymere erhalten den Band automatisch im Rahmen ihres Abonnements mit 20% Nachlaß.
Die Zeitschrift wird automatisch zur Fortsetzung weitergeliefert, sofern nicht vier Wochen vor Jahresende eine Abbestellung vorliegt.

Subscription rate of this volume:
DM 28,– plus postage.
Subscribers to Kolloid-Zeitschrift & Zeitschrift für Polymere will receive this volume additionally with a 20% discount.
The subscription will be extended automatically for unless there is a cancellation received four weeks before the end of each year.

Photokopier-Wertmarken:
Für jedes Photokopierblatt eines Beitrages oder Beitragsteiles aus dieser Zeitschrift ist eine Wertmarke von DM –,40 zu verwenden, erhältlich bei der Inkassostelle für urheberrechtliche Vervielfältigungsgebühren GmbH., 6000 Frankfurt a. M. 1.

Photostat-Stamps:
Each photostat-sheet of an article or part of an article published in this journal must show a stamp of DM –,40, which may be obtained by the Inkassostelle für urheberrechtliche Vervielfältigungsgebühren GmbH., 6000 Frankfurt a. M. 1.

Verleger: **Publisher:**
Dr. Dietrich Steinkopff Verlag, 6100 Darmstadt, Postfach 1008, Telefon (Phone): (06151) 21766 – Postscheckkonto (Postal Account) Frankfurt a. M. 95697 – Bank (Bankers): Deutsche Bank Darmstadt No. 260117. Foreign subscribers are advised to pay by cheque.

Anzeigenverwaltung: **Advertising Manager:**
Dr. Karl Niedermeyer Nachf., 6000 Frankfurt a. M.-90, Georg-Speyer-Str. 76, Telefon (Phone): (0611) 775036
Postscheckkonto (Postal Account): Frankfurt a. M. 19383.

Titel-Abkürzung: **Abbreviation of Title:**
Progr. Colloid & Polymer Sci.

Printed in Germany

GPSR Compliance

The European Union's (EU) General Product Safety Regulation (GPSR) is a set of rules that requires consumer products to be safe and our obligations to ensure this.

If you have any concerns about our products, you can contact us on ProductSafety@springernature.com

In case Publisher is established outside the EU, the EU authorized representative is:

Springer Nature Customer Service Center GmbH
Europaplatz 3
69115 Heidelberg, Germany

Batch number: 10578403

Printed by Printforce, the Netherlands